CONNECTING *Civilization*
The Growth of Communication

By Dennis Karwatka

...ections Books
...ublications, Inc.

P.O. Box 8623, Ann Arbor, MI 48107-8623

Library of Congress Control Number: 20022112855
ISBN 0-9703398-4-4
Design by Christine Ecarius

Printed in the United States of America

This book is dedicated to Philo Farnsworth, who, as a teenager, had an intuitive grasp of electronic television transmission methods. He was a tireless worker who eventually had to release his many inventions to larger organizations. Farnsworth was an intense and likable person, and his life of discovery and innovation is an honorable model for everyone.

CONNECTING CIVILIZATION

Contents

Overview

Biographies

CONNECTING CIVILIZATION

Preface

The history of world progress over the past several centuries can be clearly seen in the advancement of technology. Talented technologists used improvements in communication technology to help people become better educated, become more knowledgeable citizens, and remain in contact with each other. Their innovations brought people closer together.

This book covers communication developments by printing, by telegraph wire, by telephone, by wireless radio and television, and by Internet methods. The evolution of the fields of electricity/electronics and computer technology, which play such important roles in modern communication, are also covered in detail. The stories you will read emphasize the lives of the people who provided our modern world with wondrous gifts. After about 1800, America was often at the forefront of technical development. That is why you will read many stories of American successes within these pages. But other countries had technical heroes as well, and their accomplishments are also included.

Book Philosophy

If you're interested in the evolution of print material, you can read about it here. If your interests lean toward photography, you'll find it within these pages. The development of radio and television is also covered in this volume. All major communication technologies are included.

Wherever possible, the topics followed the lives of the people who helped develop and improve various technologies. Technical advances are always made through the efforts of individuals. This book honors their efforts by explaining how each person helped solve a technical puzzle.

For instance, the Telegraph and Telephone chapter begins with the line-of-sight mechanical semaphores built by Claude Chappe in France. It then continues to the first commercially successful electrical telegraph system, developed by Charles Wheatstone and William Cooke in Great Britain. Next comes the more practical system built in America by Samuel Morse and Alfred Vail. And the chapter continues with others in the telegraph and telephone chain. The Biographies section provides detailed individual profiles of selected key people in all communication fields. They are reprinted from my books *Technology's Past* (1996) and *Technology's Past, Vol. 2* (1999), both published by Tech Directions Books/Prakken Publications.

▶

Photographs

Technology is an image-oriented subject and I have tried to use as many pertinent photographs as possible. Some came from the extensive files of the Library of Congress, or from private companies, or other sources. I have also included photographs that I took at technical sites throughout America and in other countries. Some images are of better quality than others, but I hope each one gives you a more complete appreciation of the technical achievements of our predecessors.

Acknowledgments

A book that both traces the evolution of specific technologies and provides detailed biographies of the key players in them was suggested by a number of readers of my *Technology's Past* books. Several colleagues and friends also encouraged it. One was Dr. Jim Lewis of London who authored *London's Lea Valley, More Secrets Revealed* (Philmore Publishers, 2001). Retired Professor of English and long-time friend Gordon Haist in South Carolina was another. So was California friend and author Joe Poyer who has written over a dozen books between *The Straits of Malacca* (Doubleday Publishers, 1968) and *The M1903 Springfield and Its Variations* (North Cape Publications, 2002).

But this book and its companion volumes, *Building Civilization: The Growth of Production* and *Moving Civilization: The Growth of Transportation,* would have never seen the light of day without the efforts of members of the Tech Directions Books staff. The talents, insights, and unfailing support of Susanne Peckham, Christine Ecarius, and Pam Moore have made this a technically competent, attractive, and pleasantly readable book.

Dennis Karwatka
Morehead, Kentucky
September 9, 2002

CONNECTING CIVILIZATION

Introduction

Christopher Columbus commanded three small sailing ships when he left the port of Palos, Spain, in 1492. Some months later, they landed at San Salvador Island, north of Cuba's eastern tip. Columbus and his crew had traveled about 4,000 miles. It took over seven months before his sponsor, Queen Isabella of Spain, learned that he had discovered a new land.

Several centuries later, in 1969, Neil Armstrong became the first person to walk on the moon. Armstrong was about 200,000 miles from the launch point in Florida, but the entire world knew of the event in a little over a second. Columbus's message took so long because it was delivered face to face. Armstrong's message went electronically. Many years and many advances in technology separated the two events. The differences in speeds of message transmission emphasizes the dramatic strides that have occurred in the field of communication.

The first forms of human communication involved sound and body language. But neither has permanence, so early people took to drawing symbols on rocks and other available surfaces. Those symbols developed into written languages. Language encouraged the use of printing with movable type in the late 1400s. Graphic images encouraged the development of photography in the 1800s.

The taming of electricity added another dimension to communication. The telegraph allowed professional operators to communicate in real time even though many miles separated them. The telephone relied on related technologies, but ordinary people could use it to communicate with voice instead of through Morse code.

Electronics came into play with radio broadcasting. (Note that the word *broadcasting* comes from the agricultural term that refers to spreading seeds over a wide area.) Instead of point-to-point communication as with the telegraph and telephone, radio transmissions reached countless people at the same time. Television expanded on radio by electronically broadcasting both sounds and images.

The late 20th century saw the emergence of a new technology that provided both point-to-point and broadcast communication: the Internet-coupled computer. Individuals use electronic mail to communicate point to point with each other. Internet web sites also allow individuals and groups to view or send broadcasted information.

Modern rapid communication helps safeguard health and property. It keeps people apprised of threatening world events. It also entertains, informs, and connects people with each other. Communication fosters cooperation at all levels and has brought individuals, communities, and countries closer to each other than at any other time in history.

CHAPTER 1
Printing and Photography

Introduction

The Chinese developed movable type for printing, but it was Johann Gutenberg who introduced it to the world in the 1400s. Though his flat-bed press produced good results, his labor-intensive method resulted in a painfully slow printing rate. The 19th century rotary presses that followed were often described as "high speed" because they were so much faster than flatbed presses. Until the 1880s, all presses required setting type by hand, which presented a major bottleneck. Ottmar Merganthaler's Linotype machine automatically set type as operators depressed the keys of a special keyboard. A portion of the July 3, 1886, issue of the *New York Tribune* was the first publication to use the new Linotype machine.

Not all businesses and industries needed the large capacity of high-speed printers. Christopher Sholes built the world's first practical typewriter in 1868. Spotting a niche market, Albert Blake Dick invented a mimeograph duplicating machine for office use in 1887. Chester Carlson envisioned a machine that could immediately print a copy of an original document, patented the first dry copier in 1938 and coined the word *xerography*.

Photography does not have a history as old as printing's. It extends back to 1822 when Joseph Niepce made the first photographic image. Louis Daguerre and others worked on improvements, and in the 1830s William Talbot invented the negative-to-positive photographic method. George Eastman opened up photography to amateurs with his 1888 Kodak, the first point-and-shoot camera.

Early photographers used brittle glass plates to carry light-sensitive emulsion. Eastman's work with plastic carriers led to the creation of the motion picture industry. The Lumiere brothers made one of the first camera/projection systems in 1895 in France. Oskar Barnack created the first still camera to use 35 mm film, the German Leica, in 1925. During the 1930s, Margaret Bourke-White showed others the poetry of images found inside factories. Edwin Land invented instant photography, selling his first Polaroid camera in 1948. Today, Americans cite photography as their favorite hobby more often than any other activity.

Printing

Printing on paper began in China over a thousand years ago. Since the Chinese had invented paper, it was logical that they might also develop printing. The Chinese first used block printing. Artisans carved reverse images on a single block of wood that matched the size of the final image. Inking the surface and pressing it against a sheet of paper produced the desired results. This labor-intensive procedure encouraged the development of reusable type.

Unlike stiffly posed daguerreotype portraits, William Talbot's outdoor photographs had a modern, candid feel to them. Talbott took this photo of people with baskets of food and flowers on the grounds of his estate in 1844.

This replica of Johann Gutenberg's mid-1400s printing press is at the Gutenberg Museum in Mainz, Germany. Daily demonstrations are conducted for visiting school groups.

The Chinese were also the first people to print text on paper with movable, or reusable, type. A block printer named Bi Sheng started the method in 1045, making clay castings from wax molds. Although this was a beginning, technology and language worked against the method's implementation on a large scale.

From a technical standpoint, printers had to handle Bi Sheng's brittle ceramic typefaces very carefully to avoid cracking them during use. The Chinese alphabet had approximately 40,000 characters, which also presented considerable problems. For example, one circular type holder from the period measured eight feet in diameter. The concept of movable type slowly traveled west, where it took root in the 15th century's information explosion. The first European to achieve significant success with movable type was Johann Gutenberg (1397?-1468), a goldsmith from west-central Germany.

Johann Gutenberg. Books sold before the mid-1400s were individually handwritten or printed from hand-carved wooden pages. In either case, average people could not afford them. The development of reusable metal type that worked with printing presses caused book prices to drop and education levels to rise. Johann Gutenberg produced the first books printed with reusable metal type, an impressive two-volume large-format Bible, in 1456. Of the 200 copies Gutenberg printed, 47 still exist.

Born in Mainz, Germany, Gutenberg trained as a goldsmith. Goldsmiths marked their work with a metal punch, which may have inspired Gutenberg's interest in printing with movable metal type. Gutenberg left no written record of his motivations or details of his personal life. Though Chinese printing methods apparently influenced him, the precise sequence of his work remains unknown.

Gutenberg began to develop a method that used interchangeable metal type in 1448. To make individual letters, he engraved a character on the end of a steel punch. With a heavy hammer blow, he struck the image into a piece of copper, a softer metal, to form a matrix. Gutenberg then placed the matrix in a special two-part mold he invented. He poured in molten lead, which solidified within seconds. The adjustable mold could handle different letter widths, wider for an "m" and narrower for an "l." On a good day, Gutenberg and his assistant could cast 600 metal characters.

Gutenberg's printing press had a large screw with a heavy wooden handle. Pulling the handle forced two horizontal plates to-

Johann Gutenberg's 1456 Bible is described as a 42-line Bible. Beautiful in appearance, it is printed in Latin with an ancient typeface.

Library of Congress

This fanciful image of a young Benjamin Franklin, sometime around 1720, operating a press was painted in 1914. The other apprentice at the right holds leather bags used to ink the type. The master, perhaps Franklin's older brother James, arranges type at the left.

gether. One plate held the inked type and the other pressed paper against the type. After printing several copies of one page, the printer removed the type and reset it for another page. Gutenberg's method produced about 30 pages per hour. It remained essentially unchanged for the next 300 years.

In keeping with the practices of his time, Gutenberg's first major effort had a religious connection. He took on the task of setting all the type for a 1,286-page Bible. The Gutenberg Bible is the only very old book immediately known to much of the general public. Historians have had difficulty identifying other books printed by Gutenberg because he never put his name on his work. To finance his printing efforts, Gutenberg borrowed so much money that he earned little for his work. Late in his life, the German government awarded him a pension. He died at about the age of 71.

The printed word quickly swept through Europe. By 1500, over 40,000 separate works had been printed, representing about 20 million books. Gutenberg's two-part mold for metal type and his flatbed printing press changed human civilization forever. With improved printing equipment, people could print books and magazines inexpensively. That made literacy more important and led to the establishment of specialized fields of knowledge.

Other Flatbed Printing Press Developments. Flatbed presses like Gutenberg's remained almost unchanged for the next several centuries. Innovators devised some design improvements, but the basic operation remained the same. There was little demand for increased printing rates because populations and literacy rates stayed relatively low.

Benjamin Franklin. America's first great scientist, Benjamin Franklin (1706-1790), got his start in professional life as a printer and often described himself as one. His Philadelphia print shop, *Pennsylvania Gazette* newspaper, and *Poor Richard's Almanack* were his most important income-producing ventures. Franklin was a witty writer who used the common grammar of ordinary people. His almanac proved an immediate success when he introduced it in 1732. It was translated into a dozen languages.

Franklin personally handled all aspects of print shop operation. Like others of the time, his press had a printing rate of about 30 pages per hour. His publications' success allowed him to retire from active work in the business in 1743. Franklin maintained a financial interest in the print shop and con-

Alois Senefelder developed lithography in Germany in 1800. This image, made in 1818, was originally reproduced with a lithographic stone.

Library of Congress

tinued to help out in it. He received about £1,000 a year from the business, which easily put him in the upper middle class.

Charles Stanhope. England's Charles Stanhope (1753-1816) made one of the most notable improvements to flatbed presses. In 1800, he introduced the first successful all-metal press. Use of metal, rather than wood, made the press stiffer and produced sharper impressions.

Of noble birth, Stanhope held the title of Third Earl of Stanhope. He spent part of his childhood in Geneva, Switzerland, and took part in British politics for much of his life. He also took an interest in technology, working with calculating machines, lenses, fireproof stucco, and other items. But the all-metal Stanhope press proved his most successful invention.

Its strong iron frame allowed for printing a sheet of paper with one pull of the press handle instead of two, as was common with wooden presses. With two operators, one to feed the paper and the other to turn the screw, production ran as high as 200 pages per hour. The Stanhope press had the same screw design as its predecessors, but it used a system of levers that increased the force between the inked type and paper. The higher pressure produced sharper images, especially in those involving fine lines. The screw also required less twisting force, which reduced the operator's effort.

George Clymer. Roughly America's equivalent of Stanhope, George Clymer (1752-1834) manufactured the first American all-metal hand press in Philadelphia. Clymer's was also the first press that did not use a screw. His 1813 Columbian press used a series of force-multiplying levers and counterweights, which further speeded production while easing the work of the operator. However, since it cost more than twice as much as wooden presses, Clymer's press was not well received. In 1818, Clymer took his business to England, where he found much greater success. Beginning in the 1840s, several dozen European companies manufactured Clymer's presses.

A similar metal press that found much success in America was the Washington press. Samuel Rust invented the most popular iron hand press in the nation in the 1820s. It had a different lever arrangement from that of the Columbian, but the major factor in its success was its light weight. Made in New York City, Washington presses had hollow columns that made them easier to transport. After the expiration of Rust's patent in 1846, other companies began manufacturing their own versions of the Washington press.

This close-up of Alois Senefelder's lithographic press shows a printing stone that has an image of Senefelder.

Alois Senefelder. In the late 18th century, most people knew about traditional printing with inked type. In 1800, Alois Senefelder (1771-1834) invented *lithography*, the process of printing from a flat limestone. The ancient Greek word *lithos* meant "stone" and *graphos* meant "writing."

Senefelder was born in Prague but raised in Mannheim, Germany. His father was an actor who worked with the state theatre. Senefelder also wanted to act, but, at the insistence of his parents, he studied law. His father died when Senefelder was only 21 and Senefelder dropped out of college to try his skills at acting. When he found that he disliked the extensive traveling involved, he became a playwright instead.

In the course of having some of his work printed, Senefelder had a disagreement with a person producing his plays. He decided to do his own printing. Since he could not afford a flatbed printing press, he began to explore other methods. He decided to try reverse writing on copper plates using an acid-resisting material. Senefelder etched the plate with acid, leaving a raised image that could be inked and printed. But while this method sounded good in theory, Senefelder could not make it work well in practice. Still, the method led directly to his discovery of lithography.

One day, when he was 24, Senefelder's mother called him from the workshop to write a laundry list for her. Impatiently, Senefelder wrote the list on a polished stone with a grease pencil. Later, instead of simply cleaning the stone, he conducted an experiment with it. He poured a weak acid over the surface and found that the grease pencil protected the limestone, while the remaining stone was eaten away. The words Senefelder had written were left standing in relief. He had found a way to make a stone take ink chemically, rather than mechanically. The full explanation of the process would not be understood until long after Senefelder's death, but the simple experiment directed him toward "stone writing," or *lithography*. After several years of refining his method, Senefelder finally took out a British patent in 1800. It proved particularly popular with artists who used it for high-quality printing of artwork. Musicians also used it to print their scores in books.

In the course of further improving his method, Senefelder found that he could use other materials in place of limestone. Though printers used zinc plates widely from the 1820s on, the term *lithography* stuck. Senefelder made a series of presses, each one better than one before it, and he improved the chemical process. Like Gutenberg, Senefelder did not prosper from his new and useful printing method. He wasted his profits on continual experimentation while searching for an elusive perfect lithographic process.

Rotary Cylinder Presses. In the 19th century, newspaper publishers could not keep up with the public's demand for informa-

Lithography originally used smooth limestone surfaces. This example shows the stone master for an image of machine tool pioneer Henry Maudslay (1771-1831).

An 1814 Friedrich Koenig rotary press. A press almost identical to this one, operated by a steam engine, was the first powered press used by a newspaper. That pioneering newspaper was *The Times* of London in November 1814.

Library of Congress

Rotary presses could print more than newspapers. This one at the 1876 Centennial in Philadelphia demonstrated wallpaper printing. Its main drive power came from a power shaft under the floor. Notice how close to potentially dangerous drive belts and rotating machinery the unwary spectators stood.

tion. More people wanted to purchase newspapers than could be accommodated by traditional printing techniques. Many innovators worked on methods that would allow the continual printing of paper as it passed under a rotating cylinder. The two who had the greatest success were Friedrich Koenig (1774-1833) in Germany and Richard Hoe (1812-1886) in America.

Friedrich Koenig. Like others, Friedrich Koenig, a bookseller and printer born in Thuringia, Germany, wanted to find a method to improve the printing rate of flatbed presses. Finding little support for his ideas in Germany, he went to London to seek sponsors to pay for constructing a rotary press.

After producing many unsatisfactory designs, Koenig patented one in 1811 that was unlike any press in existence. Koenig's press still locked inked type in a flat platen, but it positioned the paper on a hollow rotating cylinder. As the cylinder rotated, it carried a hand-fed sheet of paper that pressed against the type held in the flatbed. Koenig first used the press to print books, but its increased speed appealed to newspaper publishers. Koenig built a special steam-powered experimental rotary press for *The Times* of London. To avoid a labor dispute with workers who used hand-operated flatbed presses, the November 29, 1814, issue of *The Times* was printed in secrecy.

It printed at the rate of 1,200 pages per hour, six times faster than the production results of hand presses. But *The Times* could not persuade its workers to make the change. Koenig's press went back to printing books, which so disappointed Koenig that he returned to Germany. There, in 1783, he established a partnership with Andreas Bauer (1783-1860). The company is still in business making presses in Wurzburg, Germany, on Friedrich Koenig Strasse.

Richard Hoe. Robert Hoe (1784-1833) was the first member of his family to make great contributions to printing technology. Born in central England, he did an apprenticeship in carpentry. At the age of 19, hearing of opportunities overseas, Hoe immigrated to America. While working as a carpenter, he married Rachel Smith, whose brothers manufactured wooden flatbed printing presses in New York City. When both of Rachel's brothers died in 1823, Hoe and his wife inherited the business.

Hoe went looking for a new printing product to manufacture. He heard of the steam-powered flatbed cylinder press invented by Koenig. Hoe sent one of his best workers to London to see the presses in action. Based on his worker's reports, Hoe made some changes to Koenig's design. Since America did not have widely available steam power, Hoe designed a hand-cranked cylinder press. The *Temperance Recorder* of Albany, New York, purchased his first press in 1830. Its printing rate was about 400 pages per hour. Hoe was just starting to introduce steam-powered presses when his health failed. He transferred the business to his son Richard, just before he died at age 48.

Richard Hoe was educated in the public schools of New York City. His father brought him into the printing press factory at 15, giving him a variety of jobs to help him learn the business. Richard Hoe was the first person to successfully attach printing type to a rotating cylinder. His patented hand-cranked single-cylinder flatbed presses printed 2,000 pages per hour and were in great demand. With increased business, Hoe built more buildings and his company soon had four acres under roof.

Hoe made a huge leap forward with his 1846 patented steam-engine-powered press. He eliminated the flatbed entirely. Four impression cylinders contacted one type-carrying cylinder. At each impression cylinder, a worker fed individual sheets of paper into the press. One rotation of the type cylinder printed four sheets. Its production rate was 10,000 pages per hour. Later Hoe presses had up to 10 impression cylinders and sold for $25,000.

Hoe's cylinder presses overshadowed all others, soon replacing those used by newspapers throughout America, Europe, and Australia. The R. M. Hoe Company became the leading manufacturer of presses in the world.

William Bullock (1813-1867) found a way to print from a continuous roll of paper in 1865. He referred to his process as a *web feed,* thinking of bolts of woven fabric, which were called *webs.* Bullock's method proved particularly useful for newspapers. The Hoe family quickly arranged to use Bullock's invention with their presses. Watching over every detail of his factory, Hoe suffered from overwork. He went overseas to rest and died at the age of 75 while visiting Italy.

High-speed cylinder presses made by Hoe and others played an important role in the spread of news throughout an expanding 19th century America. In 1830, when only flatbed printing presses were in use, the number of American newspapers stood at 863. Thirty years later, during the heyday of the cylinder presses, the number of newspapers had grown to 3,725.

This Washington Press was made in 1848 by the R. Hoe Company. A practical, well designed, all metal, flatbed press, this one was briefly used by author Mark Twain.

This is a model of Richard Hoe's patented 1842 two-cylinder rotary press. Two workers fed sheets of newsprint along the top plates. The method allowed for printing two identical sheets at one time.

This 1915 Linotype used melted lead to automatically cast entire lines of text. The skilled operator used a complex keyboard and other controls. It is quite similar to the one Ottmar Merganthaler invented in 1886.

Linotype Typesetter. Cylinder presses underwent continual improvement, and, by 1891, a press could produce 72,000 eight-page newspapers per hour. Although that was a high rate, it did not include the time it took to hand set the type. As a result, daily newspapers never ran more than eight pages because typesetting took so long. In the late 1870s, New York City newspapers offered $500,000 for a machine that would save at least 25 percent of the work of hand composition. Ottmar Merganthaler's 1886 Linotype machine rose to the challenge.

Born in Hatchel, Germany, Ottmar Merganthaler (1854-1899) immigrated to America when he was 18. A cousin in Washington, D.C., offered him a job in a business that made patent models. In August 1876, shortly before he became an American citizen, Merganthaler received from an inventor plans for a kind of typewriter. The machine used lithographic techniques to transfer an entire page of text. Constructing the model brought Merganthaler into contact with the printing industry and its efforts to develop an automatic typesetting machine. More than 200 inventors over a period of 50 years had attempted to patent such a machine. Merganthaler spent all his spare time on the project.

Merganthaler worked on a technique for casting an entire line of type at once. In 1884, he made a prototype for a demonstration to a small group of potential purchasers. Long, thin brass bars with individual type molds dropped into alignment with others at the touch of a character on the 90 key keyboard. Hot metal pressed against the line to form a slug and several slugs combined to make a page. The imperfect process took two more years of refinement, but on July 3, 1886, a portion of the *New York Tribune* newspaper was set with a Merganthaler experimental typesetter. Whitelaw Reid, the *Tribune*'s publisher, saw the machine at work and exclaimed, "Ottmar, you've done it . . . a line o' type." That was the casual christening of the most potent machine of its time. The *Tribune* purchased 12 Linotype machines, and before long the first hundred were in use.

A printing boom soon followed. An eight-page newspaper cost three cents in pre-Linotype days, but the cost soon dropped to one or two cents for many more pages. Within 20 years, daily newspaper circulation in America increased from 3.6 million to 33 million. The magazine industry emerged and schools throughout the nation could now purchase inexpensive books. By 1900, there were over 8,000 Linotypes operating throughout the world.

Merganthaler grew wealthy, but he never lost interest in his invention. He devised more than 50 improvements during his next few years of life. But constant work and anxiety had undermined his health and Merganthaler died of tuberculosis in Baltimore at the age of 45.

Linotype machines dominated newspaper-printing methods until the late 1960s with the perfection of photocomposition through computer terminals. Within 20 years, many newspapers eliminated their

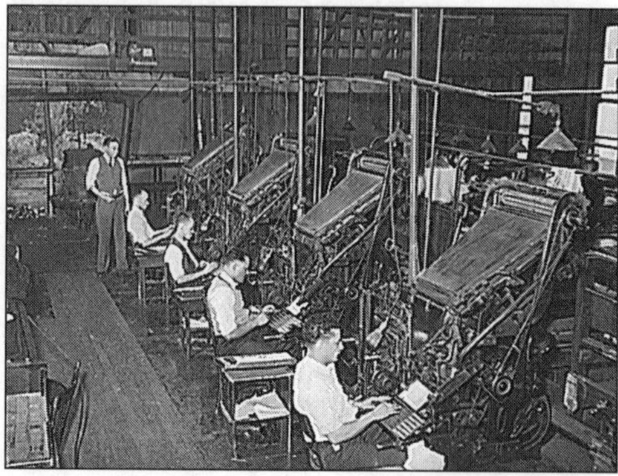

Library of Congress

Four Linotype operators set type in 1941 for the *Chicago Defender*. Founded in 1905, the *Defender* was the most influential African-American newspaper during the early and mid-20th century.

entire composing rooms. The profession had returned in some ways to where it was in Benjamin Franklin's era. Once again, the same person could do both newswriting and typesetting.

Not all printing required the high volume duplication of newspaper and book publishing. Businesses often needed just a single easily made copy of a standard letter. The first person to meet that need was Christopher Latham Sholes (1819-1890), with his practical typewriter.

Typewriter. Several inventors of the middle-1800s patented typewriters but most proved cumbersome and difficult to use. Few of the designs were human engineered and many were so tedious that potential purchasers showed no interest. Christopher Sholes led a group that developed the first modern typewriter in 1868.

Born on a farm in Mooresburg, Pennsylvania, Sholes moved with his family to Wisconsin when he was 18. He held several jobs, including serving as editor of the territorial legislature's journal. Sholes got into political life after Wisconsin became the 30th state in 1848. He served two terms as senator and one in the state assembly. President Abraham Lincoln appointed him Milwaukee Harbor Customs Collector in 1862. The job was not hard, and it gave Sholes time to experiment on typewriters. An article on the machines in *Scientific American* provided encouragement. After he discussed the article with two friends, the three men decided to try to invent a practical typewriter.

Sholes and machinist Samuel Soule rented a workspace to share with Carlos Glidden, a lawyer and part-time inventor. With Sholes as the unofficial leader, the three received a joint patent in 1868 for what is now regarded as the first modern typewriter. They also devised the four-row universal keyboard still in use on computers. It features what is often called the "qwerty" layout after the letters in the keyboard's upper left-hand corner. Sholes aimed to locate the most frequently used letter combinations as far apart as possible. This would tend to slow down a fast typist, thereby minimizing the jamming of the mechanism.

The partners may have had a good invention, but they found no ready market for it. They offered their typewriter to bankers and authors but made few sales. Bankers avoided the typewriter because they thought it would produce contracts that could be easily altered. Mark Twain purchased one, but few other writers did. No one considered the massive business possibilities of the typewriter.

The three inventors' improved 1871 model was an upright model with a horizontal rubber roller that held the paper. Sholes, the new machine's champion, tried without success to sell it. With little income coming from the invention, Soule and Glidden sold their rights to Sholes. Sholes patented additional improvements, mortgaging his home to finance his work. But the continuing lack of sales forced him to sell out in 1873 to Philo Remington for $12,000. The Remington Arms Company needed a new product to manufacture after the market for its guns dried up following end of the Civil War. The company had excellent manufacturing machinery and skilled machinists who perfected the invention. Using a well-established sales force, Remington introduced the successful $125 Remington-Sholes typewriter one year later. Sholes stayed with Remington as a consultant for a few years before he retired.

Christopher Sholes's 1871 typewriter standardized the design. It was the first to have an upright orientation, four-row qwerty keyboard, rubber paper roller, carriage return, and inked ribbon.

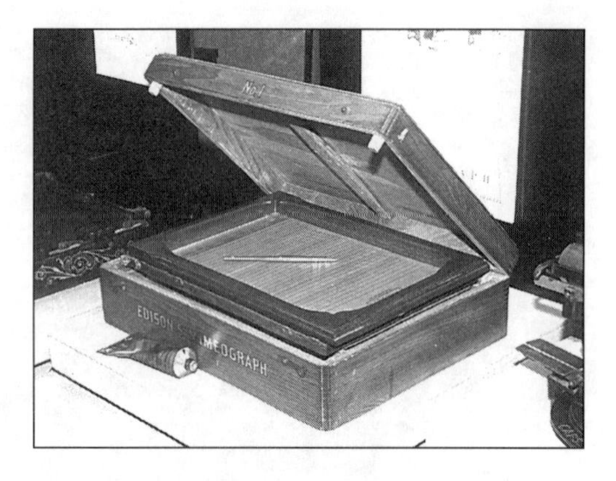

A. B. Dick patented and manufactured this single-sheet 1887 mimeograph. The photo shows the textured glass plate, stylus, and tube of ink. Dick partnered with Thomas Edison and emphasized Edison in labeling and promotion because of Edison's name recognition with the public.

Useful as the typewriter was, many businesses had a further need for several identical copies of forms or letters. A large number chose to use the mimeograph invented by Albert Blake Dick (1856-1934).

Mimeograph. In businesses and schools of the mid-20th century, the word *mimeograph* suggested a motor-operated rotating drum that made many copies of typewritten materials. However, the first mimeograph made just one copy at a time in a process similar to silk-screening. Albert Blake Dick made the first successful mimeograph machine in 1887. The ancient Greek word *mime* meant to "copy" and *graphos* referred to "writing."

Dick was born in Galesburg, Illinois, and his first job out of school was with a manufacturer of agricultural equipment. He eventually wound up in Chicago in his late 20s, where he established a successful lumber company. To communicate with customers and suppliers, Dick sent out daily inquiry sheets, typically 50 or more identical, hand-written letters. Although Christopher Sholes had recently invented a practical typewriter, business people still relied on handwriting as their basic communication method. Dick wanted to find a simple way to make multiple copies in an office environment.

Dick's solution arose from a casual experiment he conducted at his desk. He placed a wax-paper candy wrapper over a file and pulled a nail over the paper. Holding the paper up to the light, he saw that it was perforated along the line made by the nail. He reasoned that if he could force heavy ink through the holes and onto a sheet of paper, he could duplicate the line.

Dick developed textured plates, made a durable wax master, and applied for a patent. In conducting the necessary patent search, he discovered that Thomas Edison (1847-1931) had tackled a similar problem. Edison had a moderately successful patent for a vibrating electric pen that made a series of holes in a wax master. The two men met and quickly came to an agreement. Although the use of Dick's flatbed textured surface was essential to the invention, he always emphasized Edison's name on his mimeograph labels.

To use the new mimeograph, an office worker placed a stiff wax master over a sheet of specially textured glass. The person then hand wrote a message on the master with a metal-tipped stylus. Each written line or word resulted in many tiny holes. The completed master was lifted from the plate and placed on a sheet of paper. The employee rolled an inked hard-rubber roller over the master. That forced ink through the small holes and made a single copy.

In 1900, Dick developed a rotary mimeograph machine. The flatbed's frame was bent into the shape of a half-cylinder. Ink passed through the master from inside the cylinder. The cylinder rocked back and forth as an operator fed each sheet into the machine. Four years later, Dick introduced the Model 75, the first fully rotating mimeograph machine.

The mimeograph required the user to make a master by handwriting or typing. Chester Carlson (1906-1968), who worked with complicated electrical drawings, wondered if he could make instant copies from an original. His 1938 invention of xerography spawned a new technology.

Xerography. Carlson was born in Seattle into a poor family that moved often to accommodate his mother's health problems. For a while in his early teen years, he sup-

Library of Congress

This woman was photographed in 1942 using a compact, hand-operated, fully rotational mimeograph.

ported the rest of his family. He went on to put himself through the prestigious California Institute of Technology, graduating during the Great Depression. After graduation, he found work with the P. R. Mallory Company, a manufacturer of electrical components. Carlson worked long hours comparing patent drawings and text. Both often required changes, and the only methods for duplicating the complex documents were photography, redrawing, or retyping. Carlson decided to find a better way.

Carlson began his work on the dry copier by first searching the technical literature. He found that some materials, like sulfur, changed electrical conductivity after exposure to light. For his after-work experiments, Carlson rented a small room at the back of a beauty shop operated by his mother-in-law in Astoria, New York. He made his first successful experiment in 1938.

Carlson darkened the room and rubbed a sulfur-coated zinc plate with a handkerchief to develop a static electricity charge. He then pressed the sulfur plate against a glass plate that had characters written on it and exposed it to a bright light for about three seconds. He dusted the sulfur with lycopodium powder, a natural spore also known as "club moss." He gently blew on the plate, removing the loose powder and leaving a temporary image. To make it permanent, he placed a sheet of waxed paper on the powder and heated it. The waxed paper held the world's first legible dry copied image, "10-22-38 ASTORIA."

Carlson tried for five years to make a simple copying machine. All potential manufacturers rejected his invention except the Haloid Corporation, a manufacturer of photographic supplies in Rochester, New York. Carlson and Haloid reached an agreement in 1946 that gave Haloid a license to develop a copying machine based on Carlson's patents. Haloid introduced a cumbersome machine in 1949 called the Xerox Model A. Its operation required 14 separate steps and business and industry leaders did not like it. Haloid experimented with other dry models before introducing the Xerox 914 in 1960. It was the first practical dry copier. Haloid manufactured over 200,000 of the wildly successful desk-sized 914s. The company had expected to make only about 4,000 of the copiers. As a result of the 914's success, Haloid changed its name to Xerox Corporation in 1961. First used in 1947, the word *xerographic* came from the Greek *xeros* for "dry" and *graphos* for "writing."

Almost penniless in the late 1950s, after the introduction of the Xerox 914, Carlson received dozens of awards and became quite wealthy. He spent his last years giving away $100 million. Most of his anonymous donations went to research and charitable organizations.

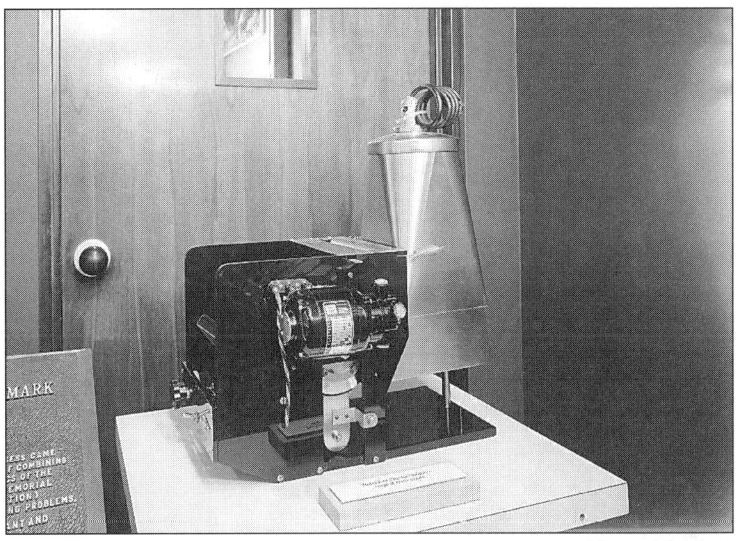

Library of Congress

This is a replica of Chester Carlson's original 1938 xerographic copier. Though the copier produced poor results, it provided a prototype for more advanced machines.

Library of Congress

The Xerox Model D copier was in production during the 1950s. It used a flat plate and required several steps by the operator. This particular copier was in use through 1985.

The first photograph with nature as its subject was made by Joseph Niepce in 1826. The indistinct image shows the courtyard at Niepce's estate. The left side shows the family's pigeon house and the top of a pear tree. The middle shows the slanting roof of the barn and on the far right is another wing of the house.

Photography

Being able to stop an image in time and record it on a piece of film has made photography a popular activity. There were many early investigators in the field, but four rise to the top of the list of pioneers:

- Joseph Niepce (1765-1833) produced the first photographic image from nature in 1826.
- Louis Daguerre (1787-1851) adapted Niepce's method in the 1830s to introduce the first practical method for taking people's pictures.
- William Talbot (1800-1877) worked on using a negative to make as many positive prints as a person wanted and in a variety of sizes. He made the first negative-to-positive print in 1835.
- George Eastman (1854-1932) made the first camera that was simple enough for the average person to use. His Kodak went into production in 1888.

Early 19th century technologists wanted to find easier ways to capture images of life on paper, glass, or metal. The earliest technique was the *camera obscura*. *Camera* meant "chamber" in Latin and *obscurus* meant "dark." The camera was a box that had a lens at one end and a 45° mirror at the other. When aimed at a building or rural scene, the mirror focused the reflected image on a piece of ground glass on the camera's top. A person would place a piece of thin paper on the glass and trace the subject.

But the camera obscura was cumbersome and people began to wonder about the possibility of "painting" with light. German scientist Johann Schulze achieved an important milestone in 1727 when he discovered that silver salts turn dark when exposed to light. Niepce used that discovery to photograph the world's first crude images in the 1820s.

Joseph Niepce. Born into a wealthy French family, Joseph Niepce trained early on for a life of religious service. When the French Revolution (1789-1799) intervened, he became an infantry officer. Returning to his home in Chalon, he married, and then formed a partnership with his older brother Claude. The brothers shared a large enough family fortune to allow them to devote much of their time to their technical interest. The brothers worked on internal combustion engines, hydraulic rams, indigo dye, and lithography, along with other pursuits.

The craze for the newly invented lithographic printing method particularly interested Joseph Niepce. His son Isidore made original drawings on stone while Niepce attended to the chemistry associated with the process. His work led him to look further into the properties of light-sensitive materials. After Claude moved to Paris in 1816, Joseph Niepce began experimenting with paper soaked in silver chloride. He had three cameras made of different sizes and attempted to take outdoor photographs. He may have had some success, but because he feared that others might steal his process, he left little written record. His earliest prints, if there were any, have been lost to history.

Niepce's chemical work took an entirely different turn in July 1822 when he began working with bitumen of Judea. This photosensitive asphalt used in lithography operates under different principles than silver salts. Niepce decided to change methods because he could not find a fixing agent for silver salts. He made his first image with bitumen of Judea, an engraving of Pope Pius VII, that year. But the world's earliest surviving image from nature was an 1826 photograph Niepce made of his courtyard.

In a darkened room, Niepce coated an 8-inch by 6.5-inch pewter plate with a slurry of the bitumen. He positioned the plate in a simple box camera with a lens at one end, aimed it out his window, and made an exposure of about eight hours. Long exposure to light made areas of the bitumen soluble. Niepce gently washed away the exposed material with oil of lavender and turpentine, which left a crude image on the pewter plate. After rinsing and drying, he made the image permanent by etching it with weak nitric acid. It is Niepce's sole surviving photograph from nature.

Niepce brought examples of his work to the Royal Society in London and to a representative of King George IV. But nothing came of the effort, partly because Niepce could not provide specific technical details. Disappointed by the experience, he accepted an offer from Louis Daguerre to enter into a partnership. Niepce thought that the younger Daguerre had enough energy and optimism to improve his process. The two men signed an agreement in 1829, but Daguerre had no photographic experience and had misrepresented his abilities. Although he worked diligently on the project, Daguerre did not make any notable progress until after Niepce died in 1833.

Louis Daguerre. Born a few miles northwest of Paris, Daguerre grew up during the French Revolution. The revolution greatly affected the country's social programs and, as a result, Daguerre received little formal education. His father, a minor government official, apprenticed Daguerre to a draftsman. He had the good fortune to serve his apprenticeship with the chief stage designer at the Paris Opera House.

Daguerre made a living by painting scenery for many of the best-known theatres in Paris. He started his best work by painting a barely visible scene on a large piece of cotton fabric. He then carefully applied an almost transparent coat of paint. A turpentine wash that followed made it even more transparent. But none of Daguerre's early work remains. Because of the combustible nature of the materials, a fire in 1839 destroyed everything.

Daguerre saw photography as a way to satisfy his desire for wealth and publicity. Hearing of Niepce's work in 1826, he read what he could on the topic. He knew that long exposure time was a limiting factor. Though Daguerre had no photographic experience, he thought he could shorten it. He wrote a letter to Niepce suggesting a partnership and the two signed their partnership agreement in December 1829. Niepce died four years later.

Sometime around 1837 and entirely by

Daguerreotypes were commonly used for portraits. The image surface was delicate and most were stored in fabric lined cases. This image shows evidence of scratching and other damage, perhaps from moisture.

This wooden camera, which has lost its lens, was made specifically for use by William Talbot, probably in 1840. The photographer focused the camera by sliding the inner box.

accident, Daguerre made three important discoveries:

• Highly polished metal plates coated with iodine could be used to capture images.

• Mercury vapor would develop the image.

• Hyposulfite would make the image permanent.

Ignoring his professional debt to Niepce, Daguerre named his finished product a *daguerreotype*. This was the first method people could use to have their photograph taken, and daguerreotypes proved very popular. The exposure time was less than one minute in bright sunlight. Since there was no negative involved in the process, daguerreotypes could not be duplicated. Each photograph resulted in just one print. They typically came in six standard sizes, the largest about 6 by 8 inches and the smallest about 1-5/8 by 2 inches.

Like Daguerre, telegraph inventor Samuel Morse (1791-1872) started life as an artist, primarily a portrait painter. Before gaining wealth with his 1840 invention of the telegraph, Morse met Daguerre while traveling in Europe. He learned how to take daguerreotypes and opened one of America's first photographic studios in New York City in 1838.

Soon after he made photography practical, Daguerre turned over his rights to the French government, in exchange for a lifetime pension of 6,000 francs per year, a very comfortable amount at the time. Daguerre retired to a country estate in 1840, while others improved on his process. He returned to painting for personal enjoyment and died at the age of 63.

Daguerreotypes dominated photography for over 10 years. But the images Daguerre's process produced were delicate and they could not be duplicated or enlarged. Daguerre's dead-end method was put to rest by William Talbot's invention of negative-positive photography.

William Talbot. William Talbot was born near Dorchester in southern England into an aristocratic family that owned a large estate. His grades put him near the top of his class in the boarding schools he attended. He entered Trinity College at Cambridge University to study classical languages and mathematics and graduated in 1821. Because of his family's wealth, Talbot did not have to work for a living. He spent his time traveling, while studying his favorite subjects of mathematics, astronomy, optics, and photography. He published over 50 scientific papers and eventually took out 12 patents.

Talbot's friends included astronomer John Hershel (1792-1871), who later suggested the word *photography*. It comes from the Greek words meaning "light" and "writing." Talbot grew particularly interested in photography during his honeymoon in Italy, when he used the camera obscura technique to make drawings of Lake Como. A lens focused the scene onto a sheet of paper and Talbott sketched the image poorly with pen and ink. Aware that he lacked artistic ability, Talbot wondered if the image could imprint itself onto a special paper. After he worked out a method, his wife, Constance Mundy Talbot, became almost as interested in the subject as her husband and became the world's first woman photographer.

Like scientists and technologists had done for centuries, Talbot built on the work of

An original 1888 Kodak camera. The string operated the shutter and the small handle on top advanced the film.

others. He studied earlier experiments to produce his own light-sensitive materials. Talbot had ordered construction of a light-tight box that had a lens in front and a sliding back to use for focusing. In August 1835, in a completely darkened room, he brushed ordinary paper with a solution of silver and iodine compounds. Talbot then placed a sheet of light-sensitive paper inside the camera. He aimed the lens at a latticed window and opened the shutter for several minutes. Talbot processed the paper negative and then made a print from it using sunlight. This first negative-to-positive photograph ever made produced a surprisingly sharp image.

Talbot took hundreds of photographs. A book he published in 1844 titled *The Pencil of Nature* was the first book illustrated with photographs. Talbot set up a photographic studio in Reading, but his prints made from paper negatives had limited commercial success. After 1851, more detailed prints could be made using wet solutions on glass plates.

Talbot not only developed photographic technology, but he used it brilliantly. His images had a modern feel. They were often interesting candid shots of people in pleasant outdoor settings. Talbot tried to include natural elements in his pictures. Few early photographs are more compelling than the ones Talbot took in the mid-19th century.

Photography in Talbot's era was the territory of the professional. It required extensive knowledge of chemistry, as well as a lot of equipment. Amateur photography did not exist until George Eastman provided the means for ordinary people to take ordinary pictures.

George Eastman. Born near Utica, New York, George Eastman had a typical early childhood. But when his father died unexpectedly, Eastman went to work as a messenger at age 14 to support his mother and two sisters. He took his first vacation, a photographic trip to Mackinac Island, Michigan, when he was 24. The simplest photographic process at the time used wet plates of glass and required the use of a large tripod-mounted camera, dark tent, glass plates, chemicals, tanks, several plate holders, and other miscellaneous items.

Immersing himself in photography, Eastman focused on making a dry emulsion. He experimented at night in his kitchen. To that point, English innovators had done most of the work with dry emulsions. No American had ever made a significant contribution to the technology of photography until Eastman. He invented a process for mass-producing dry photographic plates when he was 25, obtaining his first patent in England. Eastman sold his patent rights for $2,500, which gave him enough money to establish a factory in Rochester, New York, and enabled him to leave his job at a local bank. His dry-plate business soon soared.

A cutaway of the mostly wooden 1888 Kodak camera. The large circular opening defined the shape of the image. The photographs were contact printed and circular. The camera was advertised as: "Price $25.00—Loaded for 100 Pictures."

While looking for a lighter and more flexible support for his emulsion, Eastman experimented with paper. But professional photographers were not impressed with his product. The paper's grain was reproduced in the print after the emulsion was removed from the paper during development. Eastman then decided to test the amateur market and invented a camera for use by the general public. Eastman literally invented amateur photography with his new Kodak camera. He was the first person to put into practice the modern approach of large-scale production at low cost for a world market. The Kodak came loaded with a roll of film for 100 photographs. Customers exposed the film and mailed the entire camera back to Eastman for processing and printing. The public loved the new camera, and sales of the Kodak made Eastman an extraordinarily wealthy man.

Eastman himself came up with the word "Kodak" as a trade name for his camera. As he explained it, "I devised the name myself.

The Birtac camera projector of 1898 was one of the first amateur motion picture devices to reach the market. An incandescent gas mantle behind the wooden projector provided the light source.

The letter K had been a favorite with me. It seemed a strong, incisive sort of letter. It became a question of trying out a great number of combinations of letters that made words starting and ending with K. The word Kodak was the result."

Eastman very generously shared his good fortune with others. On just one day in 1924, for example, he anonymously donated a total of $30 million to the Hampton Institute, the Massachusetts Institute of Technology, Tuskegee Institute, and the University of Rochester. Dental clinics in Brussels, London, Paris, Rome, and Stockholm also received his financial support. In all, Eastman gave away his entire personal fortune, estimated as between $75 and $100 million. A lifelong bachelor, he died at his home in Rochester at 77.

This 1930 Leica camera closely resembled the first Leicas, which were made just five years earlier. This model had the first screw mount for its f/3.5 Elmar lens.

Eastman was the first manufacturer to use a plastic-based film carrier. He used celluloid, which was stronger than his earlier paper backing and had a smoother surface. Celluloid had first been made in 1869 by John Wesley Hyatt (1837-1920), who used the substance to replace the ivory in billiard balls. Celluloid for film was thin, tough, and transparent. It turned out to be the ideal material for early motion picture application.

Motion Pictures. The creation of a sense of motion from intermittently projected still pictures comes from a visual phenomenon known as *persistence of vision.* It was familiar to Ptolemy (c. 90-170) of ancient Greece, who described how a spinning disk appeared uniformly colored even if just a small section of it was painted. In the late 1800s, some entrepreneurs used that effect to create a viewing device that a single person could use. For a few cents, a viewer looked into a large box, turned a crank, and saw a 15-second show. Thomas Edison's 1891 Kinetoscope was the first such device. Some cities had hundreds of them in various locations.

Thomas Edison (1847-1931) was not at first interested in projecting film images onto a screen to be viewed by several people at once. He thought that more money could be made from his Kinetoscope than from motion picture houses. So it was left to French brothers Auguste (1862-1954) and Louis (1864-1948) Lumiere to arrange the first public showing of film projection to a paying audience. It took place in 1895 at the Grand Cafe in Paris.

Thomas Edison. Although Thomas Edison's name is attached to many inventions, his able staff did much of the work involved in his successes. Edison selected William Dickson (1860-1935) to take responsibility for developing a motion picture apparatus. Dickson had skills in mechanics, photography, and electricity. He made some initial experiments in 1888 with a cylinder machine that worked like early phonographs, with the sequence of pictures appearing along a spiral. The method proved impractical.

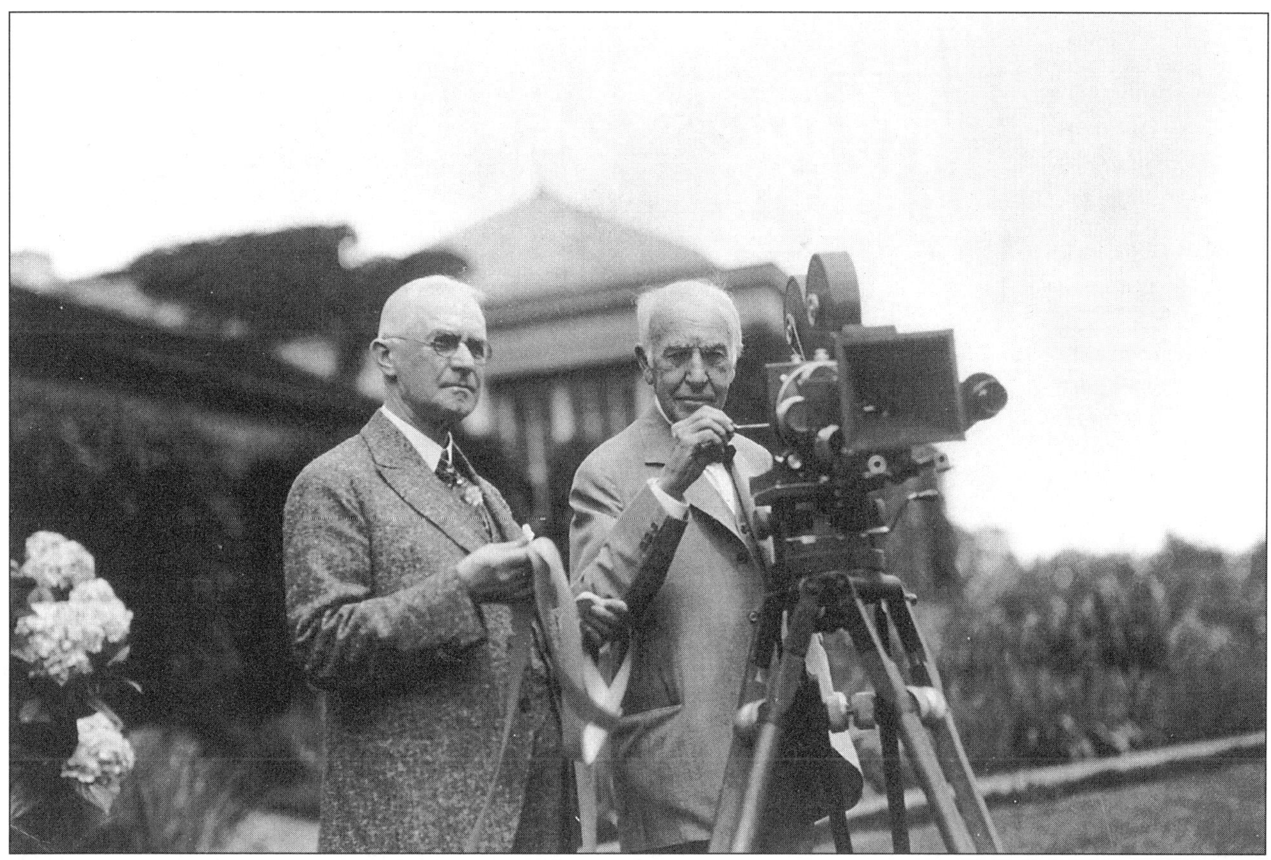

Library of Congress

Sometime during the 1920s, George Eastman (left) and Thomas Edison posed with Eastman's motion picture film and Edison's motion picture camera.

Dickson made little progress until George Eastman introduced his celluloid roll film in July 1889. Then, within two years, Dickson had working models of a camera and a viewer. The Kinetograph camera used perforated 35 mm film provided by Eastman. Camera production started in 1893, the same year that Edison set up the first film studio near his laboratory in Menlo Park, New Jersey. The studio was a small pivoted building that could be rotated to catch the sun. It also featured a hinged roof.

Kinetoscope viewers advanced long strips of film, frame by frame, in rapid succession to produce the illusion of fluid motion. The first 10 Kinetoscope parlors opened to the public in New York City in April 1894. By the end of that year, Kinetoscopes were so popular that parlors opened all over America and Europe. Although Dickson had created the first successful cinema system for Edison, the Lumiere brothers' projection method soon overshadowed it. They were the true pioneers of motion pictures.

Auguste and Louis Lumiere. Born in eastern France near the Swiss border, Auguste and Louis Lumiere were influenced by the artistic and technical skills of their father. He was a painter and an award-winning photographer. Auguste and Louis were avid photographers throughout their lives.

While attending Martiniere Technical School, Louis grew interested in dry photographic plates. Such plates had been marketed in the 1870s, but Louis developed an improved version. The brothers opened a manufacturing facility in Lyon in 1882. So popular was their product that production increased to a million a year in 1886 and to 15 million in 1894. It is impossible to determine the extent of each brother's contribution because they shared all credit on their work and patents. Many of their other relatives also worked at the factory.

The Lumieres' financial security allowed them to investigate other aspects of photography. They worked on color photography but soon put that aside after learning about the Kinetoscope. Beginning in 1894, they looked for ways to record and project motion pictures. Louis realized that the same intermittent claw mechanism that sewing machines used might be adapted to move film frames across a lens in quick succes-

Margaret Bourke-White started her career as an industrial photographer. This picture, intended for use in a newspaper article, was taken in 1940 and shows her with one of her favorite medium-format cameras.

Library of Congress

sion. He drew up crude plans and asked one of the factory's trusted technicians to construct it.

The Lumieres called their device a Cinematograph. They took it through a series of developments that made it a most remarkable device. By the summer of 1894, the Lumieres could create the negative of an image, print a film positive, and project the results at 12 frames per second. The brothers often tested their Cinematograph by filming workers leaving their plant. They showed their first projected film to the paying public in December 1895. The 10 subjects presented over a half-hour period included a baby being fed and a gardener at work. Louis had also filmed an approaching locomotive from a head-on perspective. The realistic moving image of the approaching locomotive reportedly caused some in the audience to faint and others to run from the room. Soon, people flocked to motion picture houses and the Cinematograph was in demand all over the world.

The Lumieres created films for awhile, but eventually they wanted to pursue other interests and handed their work over to others. Louis returned to research on color photography, developing the Autochrome process in 1904. Autochrome remained the favorite color method until Eastman introduced Kodachrome 30 years later. Auguste spent time investigating medical topics and served as director of a hospital's radiology department. He also wrote a medical book. Both brothers lived long lives and received many honors.

Improvements to Still Photography. Eastman was the world's largest producer of motion picture film and had standardized it at a film width of 35 mm. Companies interested in making small still cameras to compete with the Kodak sized them to use 35 mm film. One of the best known early brands was the Leica.

Eastman manufactured the Kodak camera to appeal to a large audience. Amateur photographers soon asked for small adjustable cameras for more advanced uses. Germany's Leica was among the first and best known brands.

Oskar Barnack (1879-1936) worked for Ernst Leitz's company in Wetzlar as an optical engineer. His responsibilities centered around construction of a motion picture camera. Emulsions on the films available at that time had inconsistent sensitivities to light in each batch. This resulted in exposure problems during filming. Barnack suggested making a small still camera to test short lengths of the 35 mm film for use in motion picture cameras. He built such a camera in 1913 with a sophisticated focal-plane shutter. His camera also had shutter speeds adjustable to 1/500 second, film counter to 50 exposures, and a lens with a focal length of 42 mm. The all-metal camera was the Leica prototype. Its name came from combining parts of the words "Leitz camera."

The camera's success came from having a high-quality lens fitted in a precisely machined metal body. Although Leitz had never produced cameras for the general market, the company had an excellent reputation for producing high-quality optical instruments. World War I interrupted the camera's introduction until 1925 when Leitz shipped 1,000 cameras. At the time, each one of the expensive cameras cost the equivalent of about two months' wages.

The Leica established 35 mm still photography as a practical format. Film manufacturers soon began to make the special fine-grain films that Barnack needed. He continued to work with such additional improvements as interchangeable lenses, coupled rangefinders, and 250-exposure film packets.

Improvements in still photography led newspapers and magazines to use more graphics. The photographers who took those pictures had to be part artist and part technologist. Few combined those traits better than Margaret Bourke-White (1904-1971). She was the first industrial photographer to isolate detail and record the visual drama of America at work.

Industrial Photography. Born in New York City, Margaret Bourke-White was a tall, strikingly handsome woman. She incorporated her mother's family name, Bourke, into her last name while she was in her early 20s. After graduating from Plainfield (New Jersey) High School, she earned a degree in biology from Cornell University in 1927. Her father had died while she was in college and to support herself, she photographed campus buildings and sold the prints for income. That was the casual beginning of a lifelong career that would never see Bourke-White more than an arm's length away from a camera.

After college, Bourke-White moved to Cleveland to try her luck with industrial photography. She photographed factory buildings and other aspects of the city's manufacturing community. Though not the first person to take such photographs, she was the first to give her images a moving, dramatic character. She shot her first serious industrial photographs over a five-month period at the Otis Steel Corporation.

At Otis, Bourke-White experimented with lenses, films, and magnesium flares to provide illumination. She produced a series of impressive images under challenging conditions. Although she took over 1,000 pictures with a large tripod-mounted camera, she only presented her best 12 to the company president. He liked the photographs and used them in a limited production magazine. A few months later, the Associated Press ran a national article about Bourke-White and her work.

Bourke-White photographed airplane factories, oil refineries, electrical alternators, paper mills, and countless other industrial subjects. Her images frequently appeared in *Fortune* magazine. Readers were routinely surprised to discover that the photographer was an attractive and fashionably dressed woman in her 20s.

One of *Life* magazine's original photographers, Bourke-White had responsibility for both the cover picture and lead story photos in the first issue of November 23, 1936. The cover showed three huge concrete dam supports from the Fort Peck Dam on the upper reaches of the Missouri River in Montana. When she left *Life* over 20 years later, Bourke-White had completed 284 assignments.

Early in her career, Bourke-White's images emphasized machinery, and workers played a secondary role. By the mid-1930s, the Great Depression had forced the country to pay closer attention to the problems of the general populace. Now, more and more often, Bourke-White focused her lens on the workers. In *Life*'s first issue, her cover photograph clearly showed America's technical might. However, most of her 16 photographs inside the magazine were of people—the workers constructing the dam that appeared on the cover.

Bourke-White continued to demonstrate her talent by recording dramatic emotional images during World War II and the Korean War. She twice briefly married, the second time to *Tobacco Road* author Erskine Caldwell. She died at her home in Stamford, Connecticut.

At about the time that Bourke-White's career peaked, an inventor had begun work on instant photography. Looking for a product to keep his company in business after World War II, Edwin Land (1909-1991) manufactured the Polaroid Land camera beginning in 1948.

Edwin Land invented instant photography in 1948 with his Polaroid camera. This photo shows him demonstrating a large-format version of his process.

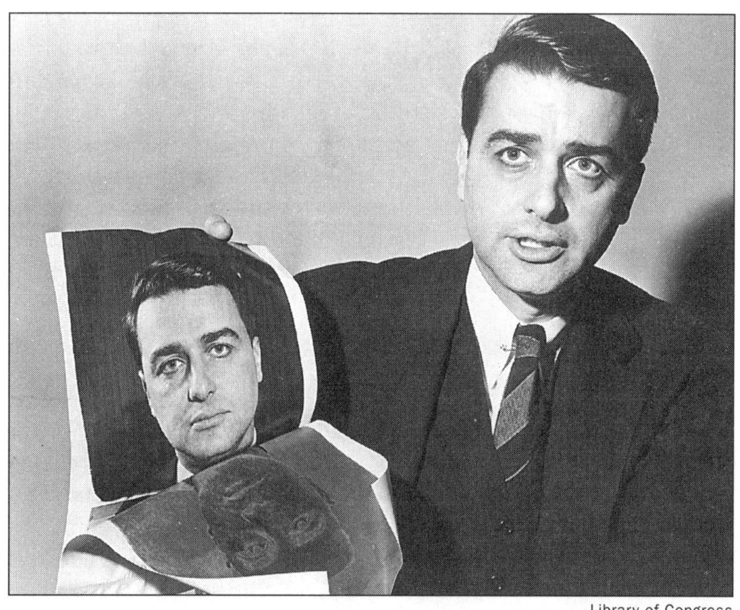

Library of Congress

The 1948 Model 95 Polaroid Land camera was the first one offered for sale to the public. A flap at the back opened and the sepia toned print peeled away from the negative strip about a minute after exposure.

Instant Photography. Land was born into an upper-middle-class family in Bridgeport, Connecticut. In 1926, he entered Harvard University, where he heard of expensive prism-type light polarizers. Light has both vertical and horizontal components, and polarizers can block either one. Polarizers held the potential for reducing glare and improving optical instruments. Land became fascinated by polarizers and considered ways to make less expensive ones.

Land dropped out of college after a year and moved to New York City. There, he spent his time doing library research and working in a small apartment on West 55th Street. A friend at Columbia University allowed Land to use the university's laboratory facilities. Forming a partnership with his friend, George Wheelwright (1903-2001), Land focused on making a polarizer in sheet form. Land and Wheelwright's general method involved using a magnetic field to align crystals of iodine while they held the crystals in place with a nitrocellulose plastic. Land applied for his first patent in 1929, the year he turned 20.

The young partners hoped to sell their polarizers for use in automobile headlights and windshields, but that application never caught on. However, the Eastman Kodak Company bought polarizers for use as camera filters. Around that time, a friend of Land's suggested he name the material *polaroid.* Land and Wheelwright established the Polaroid Corporation in Cambridge, Massachusetts in 1937. They continually improved their polarizing material, which found uses in sunglasses, variable-density windows, glareless desk lamps, and other optical products. However, none of these products sold particularly well, and Wheelwright left the company in 1942.

Land married Helen Maislen in 1929 and they had two daughters. A chance remark by one daughter, who asked why she couldn't see the photographs taken earlier in the day, set Land to working on instant photography in 1943. But he had little time then to pursue the idea. He was involved with World War II military contracts for night-vision goggles, rangefinders, tank gun sights, and many other items. His factory employed over 1,200 people.

After the war, when Land's company fell on hard times, he returned to work on instant photography. With Land providing the leadership, his scientists made a camera and special film. Boston's Jordan Marsh Department Store sold the first Model 95 Polaroid Land cameras to the public in 1948. They sold for $89.75, with an eight-exposure film pack for $1.75. The sepia-toned images took 60 seconds to develop. The new instant cameras proved the company's key to success. First-year sales exceeded $5 million, and by 1950 more than 4,000 dealers sold Polaroid Land cameras.

Land was a brilliant researcher who often worked 18-hour days in a private laboratory next to his office. He received 535 patents and was awarded the National Medal of Science in 1967. He retired from the company in 1982 and spent his final years at the Rowland Institute for Science, which he had founded in 1980. Land was a private person who did not keep a professional journal. The details of his life are fragmented partly because all his personal papers were destroyed shortly after he died.

References

Mechanisms of the Linotype and Intertype by Oscar Abel and Windsor Straw, 1961, Brookings Lebawarts Press.

Fox Talbot and the Invention of Photography by Gail Buckland, David R. Goodine Publisher, 1980.

The Gutenberg Bible by Martin Davies, British Library Press, 1996.

One Hundred Great Product Designs by Jay Doblin,

Van Nostrand Reinhold Publishers, 1970.

The Printers by Leonard Everett Fisher, Franklin Watts Publishers, 1965.

The History of Photography by Helmut Gernsheim, Thames and Hudson Publishers, 1969.

Stories of Great Craftsmen by S. H. Glenister, Books for Libraries Press, 1939 (reprinted 1970).

"Woman of Steel" by Vicki Goldberg, *American Heritage of Invention and Technology,* Spring 1987.

"Struggling to Become an Inventor" by Dean J. Golembeski, *American Heritage of Invention and Technology,* Winter 1989.

Daguerre by Beaumont Newhall, Winter House Publishers, 1971.

Printers and Printing by David Pottinger, Harvard University Press, 1941 (reprinted 1971).

The Story of Printing by Irving B. Simon, Harvey House Publishers, 1965.

"George Eastman" by O. N. Solbert, *Image* (The Journal of Photography of the George Eastman House, Inc.), November 1953.

"The Invention Nobody Wanted" by Don Wharton, *The Kiwanis Magazine,* February 1965.

A History of Invention by Trevor I. Williams, Checkmark Books, 2000.

CHAPTER 2
Electricity and Electronics

Introduction

Electricity and electronics lie at the heart of modern communications. Words heard on a radio are divided into electronic parcels that bounce off several devices before emerging from the radio's speaker. Words written for books and newspapers begin on the monitor of a computer network. Photographs commonly use digital/electronic technologies. Moving images on motion picture and television screens often result from videotaping processes.

Today, aside from its use in communication, electricity has applications in illumination, transportation, factory production, domestic settings, and many other areas of life. Alessandro Volta (1745-1827) could not have imagined those many applications when he opened the door to using electricity with his 1800 invention of the storage battery. The unit of electrical pressure, *volts*, is named for him. Others around the world added their contributions. Andre Ampere (1775-1836) used mathematics to show the relationship between electricity and magnetism. The unit of electrical flowrate, the *ampere*, is named for him. Georg Ohm (1787-1854) worked with the first electrical circuits and developed a law that relates voltage, current, and resistance. Ohm's law states that volts equal amperes multiplied by ohms. The unit of electrical resistance, the *ohm*, is named for him.

But before Italy's Volta, France's Ampere, and Germany's Ohm, came America's Benjamin Franklin. He was the first to logically investigate electricity's characteristics. Others who came later built electric motors, generators, and transmission systems. They provided the infrastructure to support an expanding electrical network.

As the 19th century merged with the 20th, electricity's close relative, electronics, began

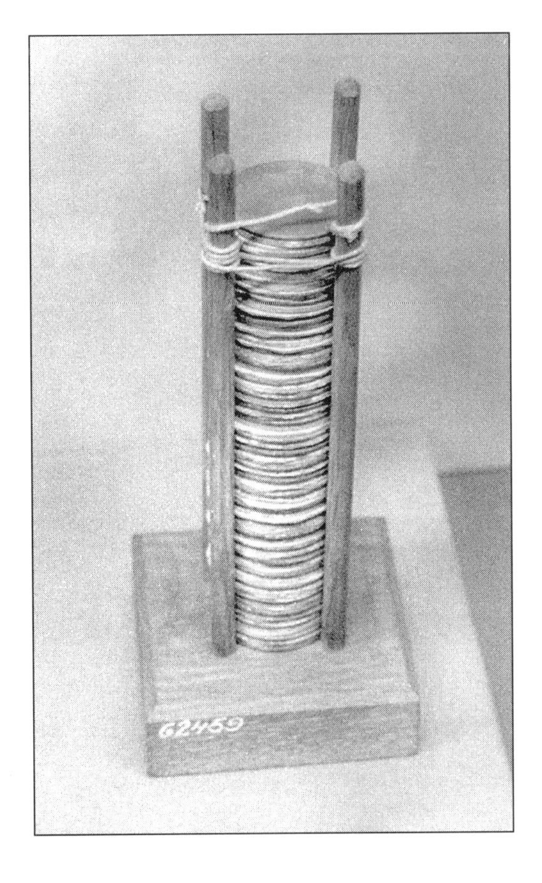

Voltaic piles were stacks of coin-sized metal discs separated by brine-soaked paper. Alessandro Volta made this one in the early 1800s. It probably developed 15 to 20 volts.

This replica of a 1904 Fleming valve resembles an incandescent lamp. The experimental device was made for Fleming by a lamp manufacturer.

Georg Ohm made this simple galvanometer in the 1820s, in the course of experimenting with the effect of resistance on current flow.

to emerge. Only someone skilled enough to interpret subtle experimental results could perceive it. Thomas Edison first observed a characteristic in 1883 that was later called the "Edison effect." But even Edison didn't know what to make of it. Britain's John Fleming and America's Lee De Forest did. Working independently around 1900, they became the earliest pioneers of electronic communication.

Benjamin Franklin often spent long periods of time in England and France. While in London, he lived in this house at 36 Craven Street, which was under restoration when the photograph was taken. It is close to the Charing Cross Railway Station.

Electronics grew in importance during the early 20th century as scientists looked for more ways to use it. The transistor emerged in late 1947, when William Shockley and his team worked on solid state amplifiers. Then came integrated circuits (ICs) in the late 1950s, invented independently by Jack Kilby and Robert Noyce. ICs allowed people to make communications devices smaller and smaller.

Electricity

Early 19th century technologists could not have predicted the importance of electricity. They saw it primarily in the form of bolts of lightning and the type of harmless shocks that result from scuffing shoes on a rug. The word *electricity* comes from the Latin *electrum*, meaning "amber." Amber is fossilized tree sap similar in composition to modern plastic. Rubbing it with fabric produced static electricity that attracted small scraps of paper and lifted animal fur.

Many people know Benjamin Franklin (1706-1790) through study of American history in school. The United States' colonial government sent Franklin to France during the Revolutionary War (1775-1781) because he was the American who commanded the greatest respect in Europe. That respect resulted from Franklin's brilliant work with electricity. He was the first of the world's electricians and his writings on the subject were translated into French, German, and Italian.

Benjamin Franklin. Born in Boston, Benjamin Franklin apprenticed at 12 to his older brother James, who started publishing *The New England Courant*. Between 1723 and 1729, Franklin worked for a number of printers before opening a Philadelphia print shop in partnership with a friend. Two years later he had become the shop's sole owner, and it remained a lifelong source of income for him.

In 1746, Franklin saw a series of electrical experiments in Boston. The mystery of electricity captivated Franklin, who went on to conduct many experiments over the next 30 years. His most dramatic experiment involved flying a kite in a Philadelphia field in 1752 as a thunderstorm approached the city. The kite had a pointed wire connected to its string. A metal key dangled from the string near Franklin's hand. Lightning crack-

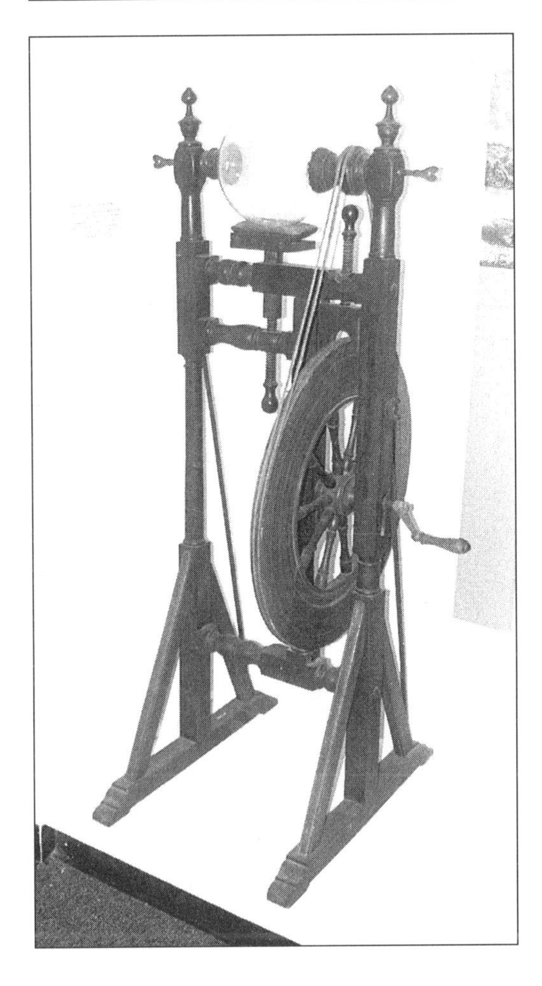

Benjamin Franklin may have used this electrostatic machine in the 1760s, but conclusive evidence is lacking. It is displayed with some parts missing.

led in the distance and a small amount of electricity traveled through the air. The pointed wire picked up the electricity and sent it down the damp string. Franklin drew a spark from the key to his hand, which proved that lightning was electricity. Franklin's dangerous experiment showed that the weak electricity produced in a laboratory was directly related to powerful natural forces. He was extremely lucky that his experiment did not injure him. Lightning electrocuted the next two people who tried to duplicate the procedure.

Franklin also operated electrostatic generators that developed high-voltage sparks. In step-by-step fashion, which was unusual at that time, Franklin showed that electricity had either a positive or a negative charge. He was the first person to investigate the details of electrical storage devices called Leyden jars, the precursors to capacitors. Franklin was also the first to describe aspects of electricity using such words as "plus" and "minus," "conductor," "charge," "battery," and "electric shock." His election to the British Royal Society in 1756 and the French Academy of Sciences in 1772 acknowledged his achievements as a leading 18th century scientist.

Franklin's electrical discoveries in the late 1700s did not inspire others to rush to study the field. But in 1800, Volta's battery kindled great interest. Pioneering work with electricity reached unimagined heights under the able lead of Britain's Michael Faraday (1791-1867). His two greatest accomplishments were inventing the electric motor in 1821 and the transformer in 1831.

Michael Faraday. The third of four children, Michael Faraday was born in a London suburb. His father, a blacksmith, suffered from poor health that kept him from regular work. The family got by with only the bare necessities of life. When he was 13, Faraday apprenticed to a sympathetic bookbinder who allowed him to read everything that passed through his shop. He liked science books, especially ones dealing with chemistry and electricity. After teaching himself both subjects, Faraday found a job as an assistant chemist at London's Royal Institution at age 21. The position allowed Faraday to develop his full potential and he remained at the Institution for the rest of his working life. The Royal Institution had been established in 1799 partly through the assistance of an American Benjamin Thompson (1753-1814), also known as

Library of Congress

Many photos were taken of Michael Faraday. This image is a damaged daguerreotype taken around 1850.

Michael Faraday's 19th century laboratory is displayed behind glass at the Royal Institution. A large electromagnet rests on the low chair under a table at left center.

Count Rumford, who helped finance the purchase of a large building for £4,850.

Early on, electricity was viewed as a branch of chemistry, and much of Faraday's chemical work used batteries. The recently discovered and mysterious force confused many investigators, but not Faraday. He made a device in 1821 that used a battery, a magnet, a dish, a small amount of mercury, and a wire hanging from a pivot. Once set in motion, the wire continued rotating around the magnet. This very first electric motor showed that electricity could be used as a source of mechanical power. Faraday's laboratory skills were unmatched and, in 1825, he became head of the Royal Institution. He continued to operate his small laboratory in the basement and lived with his wife in an apartment on the third floor.

Faraday succeeded at using electromagnetism to induce a flow of electricity in 1831. He bent an iron bar into a 6-inch-diameter ring and wrapped the left side with many turns of wire. He then wrapped the right side with wire and connected the wire terminals from one side of the ring to a galvanometer. When the wire terminals from the other side of the ring were connected to a battery, the galvanometer needle swung toward full scale. Disconnecting the battery caused the galvanometer to swing in the other direction. Faraday found that a current change in one coil produced a current flow in the other. This was the world's first transformer and many consider it Faraday's greatest contribution to technology.

Faraday proved a tireless, and often successful, worker. He made an electrical generator in 1831 that essentially consisted of a copper disc that rotated between the poles of a horseshoe magnet. Faraday also developed the idea of magnetic fields and lines of force. He discovered the effect of magnetic forces on light beams. He designed a device for measuring electric charge. The list goes on and on.

Faraday had a calm, gentle manner and showed little interest in money or personal recognition. While others worked to achieve a "first" in some aspect of electricity, Faraday created what he called his Christmas Lectures for young people. Beginning in 1826, he dazzled young audiences with technical experiments and colorful stories. The Christmas Lectures, now televised, still continue at the Royal Institution. At the age of 67, Faraday retired to a house provided by Britain's Queen Victoria. He died nine years later. Faraday is the only person honored by having two measurement units named after him. The *farad* is a unit of capacitance and the *faraday* is a unit of charge.

People who work with new ideas in technology often fall into one of two classifications. Those working with hardware are called *experimental investigators*. Faraday was among the best experimentalists the world ever produced. Those working primarily with pencil and paper are called *theoretical investigators*. Charles Proteus Steinmetz (1865-1923) was a theoretician who built on

This demonstration model of Michael Faraday's generator was made in the late 19th century. His generator was the first device to convert mechanical motion into direct current.

General Electric Company

Faraday's discoveries. More than any other person, the German-born Steinmetz had responsibility for establishing America's early leadership in the field of electricity.

Charles Proteus Steinmetz. Although of German heritage, Charles Proteus Steinmetz was born in a city that is now Wroclaw, Poland. His father headed the printing department at the local railroad office. Steinmetz's mother died during his first year and his grandmother raised him. An outstanding student at the University of Breslau, he specialized during his last year in astronomy, chemistry, electrical engineering, mathematics, medicine, and physics.

Steinmetz also associated with the student branch of the Social Revolution Party. Its political activities came under official suspicion. Shortly after completing his university studies, Steinmetz hurriedly left home on the morning he would have been arrested. He was 24 years old and he never saw his family again.

Steinmetz immigrated to America and found employment in Yonkers, New York, with Rudolph Eickemeyer, who was developing electrical equipment that used alternating current. Steinmetz first took on the problem of *hysteresis*, the loss of efficiency caused by alternating magnetic effects. Many scientists did not believe hysteresis existed because at the time no one had measured it. But using existing data and high-level mathematics, Steinmetz proved the existence of hysteresis. He also showed a method for reducing it. When he presented his results to the American Institute of Electrical Engineers in 1892, Steinmetz became an immediate sensation in the technical world.

The newly formed General Electric Company (GE) wanted Steinmetz to join its staff. He refused, not wanting to turn his back on the employer who gave him his start in America. As a result, GE offered to purchase Eickemeyer's company, with the expectation of receiving Steinmetz's services. Everyone agreed and Steinmetz moved to Schenectady, where he remained for the rest of his life.

Electrical genius Charles Steinmetz overcame a birth deformity and helped establish America's early dominance with alternating current equipment. According to his contemporaries, Steinmetz was fond of smoking foul-smelling cigars.

Soon after his arrival in 1892, GE established the first company-sponsored industrial research facility. Still in his 20s, Steinmetz was made consulting engineer in charge of advanced theoretical analysis. He described one workday in writing, "It was a hot sunny day with almost no wind, and I sat in the sun and calculated instances of condenser discharge through an asymmetrical gas circuit." Steinmetz received 195 patents, published 10 technical books, and gradually made alternating current electricity less confusing. He showed the advantages of applying mathematical methods to practical electrical problems.

As with his contemporary Albert Einstein (1879-1955), the public had no understanding of Steinmetz's work. But people recognized him as America's foremost electrical genius. Steinmetz's three most important achievements were his:

- Mathematical investigation of magnetism;
- Development of a practical method for making calculations in alternating current systems; and

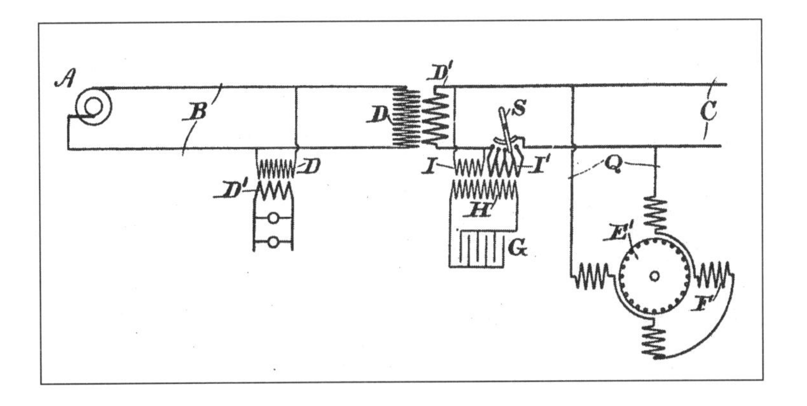

The patents that Charles Steinmetz took out were often based on sophisticated theoretical mathematics. This one from 1894, which shows a method for preventing phase displacement in alternating circuits, is no exception.

• General study and theory of electrical transients.

Steinmetz loved children but had no family of his own. In 1905, he legally adopted the family of his lab assistant, Joseph Hayden. Hayden, his wife, and their three children shared Steinmetz's large rambling home. Steinmetz died of heart failure shortly after returning with the Haydens from a tour of the West Coast, his first long vacation trip.

Other 19th Century Pioneers. The development of electricity into the early 20th century followed no specific pattern. It wasn't like printing, which evolved in an almost linear way from flatbed presses, to rotary presses, and then to the Linotype. Nor was it like photography, which also had an almost linear development from daguerreotypes, to negative/positive methods, and then to roll film for still and motion picture cameras.

Many early workers in electricity followed their own paths. Some wanted to apply electricity to telegraph or telephone communication. Others worked on using it for illumination with arc lamps or incandescent lamps. A few focused on its use in transportation with streetcars, automobiles, or locomotives. Potentially powerful electric motors could drive individual factory machine tools, replacing the single overhead line shaft. Electricity also held promise for making domestic tasks easier through use of electric refrigerators, stoves, washing machines, and other appliances. The different paths seemed almost limitless.

Joseph Henry was an early electrical pioneer who became the first director of the Smithsonian Institution in 1846. His bronze statue is in front of the Smithsonian's first building, which people call the Castle.

Because of the varied and disjointed evolution of electricity's use, it would be impossible to mention all the people who played important early roles. Countless technical heroes made important contributions, but the few who follow rise to the top.

Joseph Henry. Born in Albany, New York, Joseph Henry (1797-1878) came from a poor family. He was involved in acting as a youth, but his interests changed when he began reading popular science books. After working his way through college, Henry took a teaching position in 1826 at the Albany Academy, where he began his electrical work. He was the first American to experiment with electricity in any important way since Benjamin Franklin almost 80 years earlier.

Using homemade batteries, Henry made electromagnets from various lengths and diameters of wire that he wrapped around iron cores. He found that winding the wire in sections allowed him to join them in different combinations. By 1831, he had made an electromagnet that could lift 750 pounds. Princeton University rewarded him with a professorship.

Henry may have independently discovered the principle of electromagnetic induction before Faraday. But Faraday published his results first and received credit for the discovery. Still, there were no hard feelings between the two men, who met in 1835 when Henry visited England. Faraday proposed that the Royal Society of London, the world's most prestigious scientific organization, award Henry its coveted Copley Medal. Only one other American, Benjamin Franklin, had been so honored.

Henry moved to Washington, D.C., in 1846 to serve as the first director of the Smithsonian Institution. The fledgling organization was searching for a purpose and direction. Henry resisted popular ideas, such as focusing efforts on a school of steam engineering or an extensive lecture series. He decided to use the Smithsonian to distribute information among all the people working in technology. Some call that his most valuable invention. Henry remained at the Smithsonian until his death. The unit of elec-

Library of Congress

George Westinghouse had more responsibility for introducing alternating current electricity to the public than any other person. This photograph of him was taken in the early 1900s.

trical inductance, the *henry*, is named for him.

George Westinghouse. At just 22 years of age, George Westinghouse (1846-1914) got his start in technology by inventing an air brake for trains. His invention met with such immediate demand that he formed the Westinghouse Air Brake Company in 1869 with a capitalization of $500,000. The company provided financial support for Westinghouse to work on inventions. He made contributions to railroad signaling systems, natural gas systems, turbines, and electrical equipment that used alternating current. In all, he held over 350 patents.

After his air brake, Westinghouse's most important contribution was introducing alternating current electricity for public use. Westinghouse knew that ac was easily changed in transformers and that it could be sent over high-voltage transmission wires more efficiently than direct current. His company developed ac equipment and steam turbines to power the alternators. He purchased Nikola Tesla's ac motor patents and hired Tesla to work on development.

At the time, Thomas Edison advocated the use of direct current electricity because it was easily stored in batteries. Edison sold only dc-generating equipment. Because of their difference of opinion, Edison and Westinghouse were rivals for the huge consumer electrical market. Following a competition described by newspapers of the day as "The Battle of the Currents," Westinghouse won a crucial victory over Edison. He received the contract to install three massive 5,000 hp ac alternators at Niagara Falls. The alternators went on line in November 1896, delivering power to Buffalo, 22 miles away.

Westinghouse acquired wealth at an early age. In most ways, he lived modestly, though he did build a large home in Pittsburgh for himself, his wife, and their only child. He also owned his own railroad car. Westinghouse was a private man who kept no diary, wrote few letters, and refused to be interviewed by reporters.

Nikola Tesla. Nikola Tesla's (1856-1943) greatest achievement can be summed up in six words: He made alternating current electricity practical. Tesla did this by designing practical ac motors, alternators, and transmission equipment. Born in Croatia, Tesla showed remarkable talent in college. After graduating, he held some minor jobs in Europe, then immigrated to America in 1884 to pursue more challenging opportunities.

A glowing letter of introduction got him a position with Thomas Edison at Edison's Pearl Street electrical-generating plant in New York City. Tesla began by working on shorter pole pieces for direct current dynamos. Later, he and Edison had a disagreement, and Tesla quit on the spot. His feelings ran so deep that when, in 1912, Tesla was offered the Nobel Prize in physics to be shared with Edison, he refused the honor.

Tesla worked for several electrical companies, while investigating alternators and ac motors in his spare time. He received several patents in 1888. Meanwhile, George Westinghouse, who was committed to ac, was having problems developing suitable motors and alternators. Tesla's patents filled the void and the two men agreed to work together.

Underbidding Edison, Westinghouse won the contract to illuminate the 1893 world's fair in Chicago. Tesla assembled 24 500 hp alternators to provide electricity. His equipment originally operated at 133 cycles per second, but he found that to be too high. Tesla experimented until he came up with 60 cycles per second. That frequency proved both low enough to operate motors and high enough to eliminate lamp flickering.

A model of the four-story 1882 lighting plant that Thomas Edison operated in New York City. The ground floor has four boilers. The second floor has six steam-powered dynamos and the upper two floors housed distribution equipment.

Then, Westinghouse won the contract to build alternators to harness the power of Niagara Falls. The project produced the world's first major hydroelectric generating plant. It used three huge water turbines connected to Tesla-designed two-phase alternators. The power plant went on line in 1896. When completely operational in 1902, that plant alone produced more electrical power than that produced in 31 states combined. All the plant's generation and transmission systems came from Tesla's patents.

Though a technical genius, Tesla was arrogant and tactless and, as a result, he had no friends. People found it almost impossible to avoid offending him. He died alone at 86 in his New York City apartment. The *tesla coil*, used in high-voltage, high-frequency electrical demonstrations, is named after him.

Thomas Edison. After Thomas Edison (1847-1931) operated his successful incandescent lamp in 1879, he faced the monumental problem of producing and distributing electricity to customers. In an old New York City warehouse, he installed six steam-powered dc dynamos and delivered electrical power through buried conductors. The Pearl Street station was the world's first permanent central electric power plant. It went on-line in September 1882.

Early electric plants only provided electricity for lighting. The brightness of early incandescent electric lamps was 16 candlepower (cp), which matched the output of natural gas lamps. A modern 12-watt lamp would produce about 16 cp. Each dynamo could provide power for 1,200 lamps. But within 14 months of opening, the Pearl Street station's six dynamos were lighting up to 10,664 lamps, far more than the 7,200-lamp design capacity. Edison had to continually upgrade his equipment. When asked why he chose 110 volts as the standard for his lamps, Edison said: "I based my judgement on the best I thought we could do in the matter of reducing the cost of copper and the difficulties we had in making filaments stable at high voltages. I thought 110 volts would be sufficient to insure the commercial introduction of the system."

Edison's four-story building at 257 Pearl Street had four Babcock and Wilcox steam boilers on the ground floor. Six steam engine/dynamo power units were placed on the strengthened second floor. Edison put manually operated electrical distribution controls on the third floor and about 1,000 lamps for load testing on the fourth floor. To deliver the power, Edison's employees buried 15 miles of wire in trenches.

The light plant served an area of about one square mile on Manhattan's lower east side, with boundries shrewdly chosen. The area included the New York Stock Exchange, several large banks, and the offices of *The New York Times* and other influential newspapers. Edison supplied electricity free for

This dynamo is the only one that survived a fire at Edison's New York City light plant in 1890. Named Jumbo Generator Number 9, it was declared a National Historic Mechanical Engineering Landmark in 1980.

Edward Weston provided this patent model for an 1878 dynamo. The U.S. Patent Office no longer requires models.

the first three months because he could not accurately measure individual usage. His first electric bill, totaling $50.40, was delivered to the Ansonia Brass and Copper Company in January 1883. Within a year, the number of customers had risen from an original 59 to 508. In January 1890, a short circuit caused a fire that destroyed the Pearl Street plant.

Making electrical measurements through the late 1800s was not a simple matter. The problem so distressed Edison that he devised his own method for use at his light plant. Edison's earliest ammeter was a nail hanging from a string positioned near the electrical supply line. Different current flows caused magnetic effects that pulled the nail near the wire. Edison made load adjustments based on this crude measurement.

The most common meters were delicate galvanometers, which made only comparative readings. Galvanometers could not be moved during a series of experiments because the earth's magnetic field affected them. Thunderstorms—and even the nails in shoes—would upset meter readings.

Edward Weston (1850-1936), a successful manufacturer of dynamos, decided to build a voltmeter after conducting a week-long electrical experiment at the Franklin Institute's well-equipped dynamo laboratory in Philadelphia. He had expected the experiment to take only one day. He took time out from his busy life in 1886 to build the world's first practical voltmeter.

Edward Weston. Born in Shropshire, England, the young Edward Weston conducted electricity experiments at home with glass jars, sulfuric acid, and small pieces of copper and carbon. He went on to study medicine and was near the end of a medical apprenticeship when, during a train trip, he met an American tourist whose conversation inspired him to immigrate to America in 1870.

Weston found employment in New York City with a nickel-plating company that obtained voltage for plating from troublesome batteries. The batteries provided only a thin and poorly bound plating of nickel. Weston developed a dynamo that produced a better-regulated and higher current than the batteries. His work led to the first-ever use of a dynamo for electroplating. The patent he took out at age 22 formed the basis for a successful electroplating partnership that he established. Income from the patent allowed Weston to continue his work on dynamos and he soon developed one for arc lighting that proved far superior to its competition.

To make a practical voltmeter, Weston expanded on an 1881 French patent for an unsuccessful meter. It used a small coil of wire suspended by a filament inside a magnetic field. A dial connected to the coil moved over a calibrated scale. The design is still called a *D'Arsonval meter* movement after Jacques Arsene D'Arsonval (1851-1940).

Weston spent two years improving his design before he started production in 1888 at a large factory in Newark, New Jersey. The 100 volt dc meter he called the Model One was the first truly portable voltmeter

This early Weston ammeter was made in the 1890s. The moving coil can be seen under the circular glass cover at the bottom.

Ohio-born Granville Woods had so many electrical inventions, that the governor of Ohio in 1974 officially declared him "the Black Edison."

ever made. It proved an immediate success with colleges, power companies, and electric equipment manufacturers. Each sold for about $70 and was accurate to 1/4 percent. Weston invented the shunt resistor for ammeters and soon made ammeters, along with ohmmeters and wattmeters. He received over 300 patents, many for electrical meters. He died at 86 while cruising on his yacht in Long Island Sound.

People like Westinghouse, Edison, and Weston were hard workers who had good ideas. Their names are familiar to many modern technologists. Other electrical pioneers were also hard workers with good ideas, but major success eluded them. One example was Granville Woods (1856-1910). He received over 50 patents for such inventions as improved electric motors, electric transportation equipment, and electrical communication methods.

Granville Woods. Born in Columbus, Ohio, Granville Woods attended elementary school only briefly. He started working in a machine shop at the age of 10 and took private tutoring classes in the evening. Leaving home at 16, he moved to Missouri to work as a fireman and then as a locomotive driver. Woods liked all aspects of railroad work, but he developed a particular interest in telegraphy. Telegraph lines followed the rail lines, and Woods made an effort to learn all he could about electricity and electrical transmission of information.

At the age of 28, Woods used money he had saved from various jobs to establish a business with his brother Lyates. Their Cincinnati machine shop repaired pumps and measuring instruments and made gears and tools. Woods worked on electrical devices in his spare time and took out a patent in 1884 for a telephone transmitter that used a flexible diaphragm and carbon. In selling the patent to Boston's Bell Telephone Company, he did not require that Bell use his name with the transmitter. That decision was probably a mistake, and it set a lifelong pattern. It is the main reason that he is not well remembered.

Woods had his best year in 1887, when he received seven patents. He designed a regulator for electric motors that allowed the user to vary speed without using power-robbing resistors. The regulator worked so well that it produced an energy savings of 40 percent. Woods's most sophisticated invention was the induction telegraph, a system that permitted communication between a moving train and a railroad station.

Woods reorganized his company to begin manufacturing his own inventions. However, lacking the proper tooling, he went back to selling his patents. Woods wanted to make electricity practical for transportation and many of his patents reflect that interest. In 1888, he worked on a system of elec-

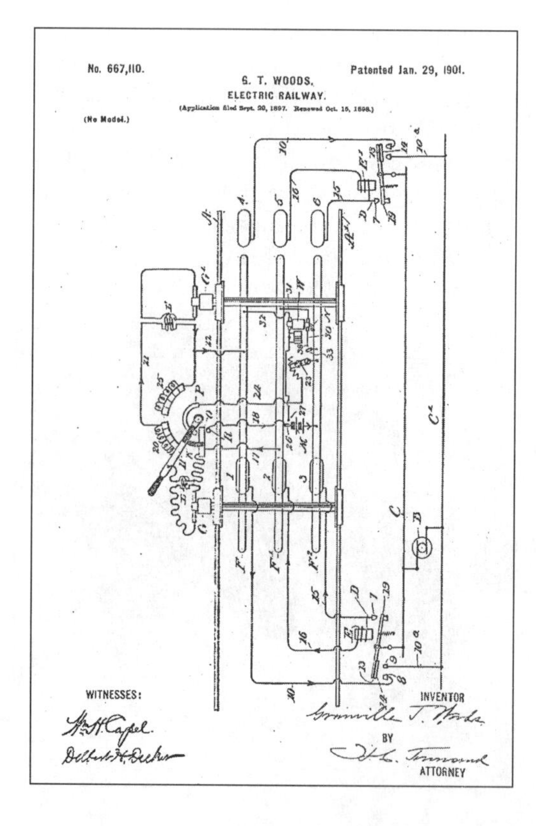

Granville Woods had many electrical patents to his credit. This 1901 patent is for electrically driven railroad cars. The car's wheels are shown rolling along the outer rails.

tric streetcars for mass transportation. The cars drew their power from an overhead line. One of Woods's inventions that continues into the 21st century is the third-rail power pickup on subways and urban commuter trains. Subway cars still draw electricity from a protected power rail outside the two rolling tracks. Woods received patents for electromagnetically controlled air brakes and automatic circuit breakers for operator safety on trains. Around 1890, he sold his share of the business and moved to New York City. He died of a stroke at 53.

Like Edison, Woods devoted his career to invention. Also like Edison, his inventions came primarily in the field of electricity. In an October 1974 proclamation, Ohio governor John Gilligan honored the African-American Granville Woods by describing him as "the Black Edison."

Electronics

Electricity and electronics are related fields, but with some significant differences.

• Electricity deals with fairly large current flows moving down a wire. It is a method for delivering power. Although exceptions exist, electrical current flows are typically 1 ampere or more.

• Electronics deals with transmission of electromagnetic signals. It also includes amplification of a signal using vacuum tubes or solid-state devices. Although exceptions exist here, too, electronic current flows are typically in the milliampere (1/1,000 ampere) range.

Thomas Edison was the first person to observe an electronic characteristic. He did so in the course of trying to improve the carbon filaments in his incandescent lamps in 1888. The filaments had a short life and tended to darken the bulb's glass globe. Edison tried to improve bulb performance by placing an extra electrode next to the glowing filament, hoping that it would absorb excess carbon.

Edison observed that when he connected the extra electrode to a positive voltage, a small current flowed between it and the filament. This was later called the *Edison effect.* Edison saw no special application for the effect but patented it anyway. That phenomenon was the key to the first radio tube. Intelligent as Edison was, he did not appreciate the importance of the subtle secret he saw. A few years later, John Fleming (1849-1945) correctly viewed the Edison effect for what it was: introductory electronics. Fleming created the field of electronics with his 1904 invention of the first practical radio tube.

John Fleming. Born in Lancaster, England, John Fleming grew up in London. His father was a minister and his mother came from a family that had developed the manufacture of portland cement. His family was comfortable, though not wealthy. Fleming graduated from the University College of London (UCL) in 1870 and went to work in Cambridge University's laboratories. He also worked at the Edison Electric Light Company in London before accepting a professorship at UCL, where he remained for 41 years.

Like Edison and others, Fleming wanted to improve the carbon filaments in incandescent lamps. In 1889, he had special three-terminal lamps made up for experimentation and began serious research. Fleming was the first person to work in the field of applied electronics. He experimented with electrical conduction from glowing filaments in a vacuum. Wanting to achieve wireless communication, he carried out countless experiments on transmission and reception. He had limited success.

Fleming was a friend of Italian radio developer Guglielmo Marconi (1874-1937). He became a technical advisor for Marconi's company in 1899 and helped Marconi design a powerful wireless transmitter in southwestern England. That transmitter sent the first faint transatlantic radio signal to St. John's, Newfoundland, in 1901.

The signal was hard to detect and Fleming wanted to improve the radio circuitry involved. He faced the challenge of converting a weak alternating current into a direct current to operate a receiver. Fleming realized that his 1889 three-terminal lamp could convert ac to dc because it let current flow in only one direction. He had correctly evaluated the Edison effect. Then, Fleming made up a small circuit with one of the lamps. In 1904, he experimentally confirmed that his invention could detect radio waves. Fleming's discovery marked the beginning of the electronic era.

Incoming electromagnetic signals have positive and negative aspects that balance each other out. People cannot hear, or de-

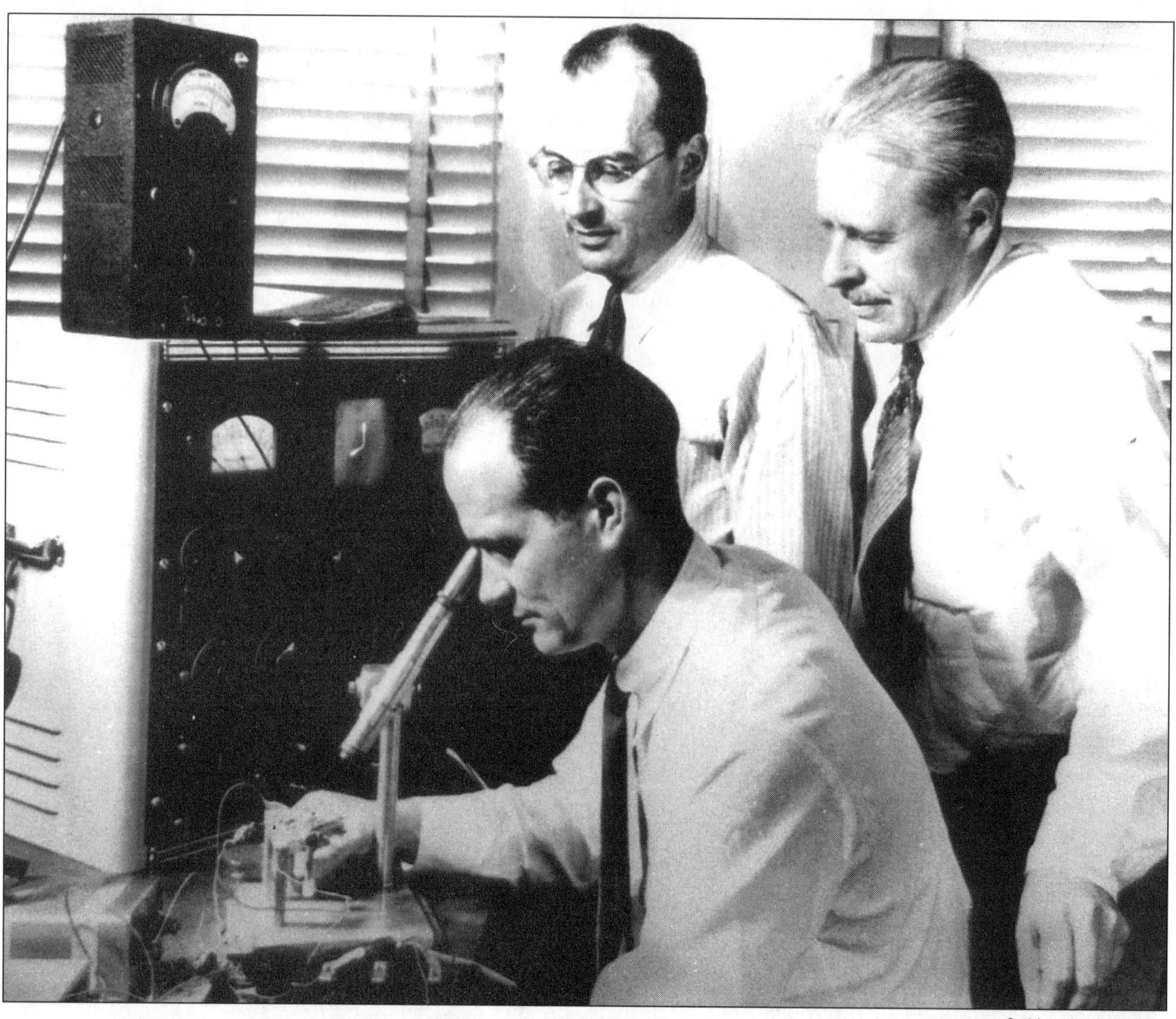

Smithsonian Institution

William Shockley (front), John Bardeen (left rear), and Walter Brattain posed for a photograph depicting the 1947 experiment that produced the transistor.

tect, such signals with earphones. Fleming's tube-type diode allowed current to flow in only one direction. It essentially wiped out the negative part of a signal, which allowed more precise detection of radio waves. Fleming's diode was heavily used in the early years of amplitude modulated (AM) radio communication. It was an essential part of all transmitters and receivers for over 50 years.

Fleming called his diode a *valve,* since it turned on when current flowed in one direction and turned off for flow in the other direction. It worked just like a one-way, or check, valve in hydraulics. Britons still call the tubes "valves," but they are known as *vacuum tubes* in America.

Fleming married twice but had no children. He was a captivating speaker who made his last public presentation at the age of 90. He lived long enough to see his invention used in advanced communication systems, such as television. A remarkably energetic individual, he remained professionally active through out most of his life. He died at the age of 95.

Fleming's valve ushered in the field of practical electronic communication. But it was only a detector, not an amplifier. It laid the groundwork for the triode amplifier invented by America's Lee De Forest (1873-1961) in 1907. The triode made it possible to receive voice communication and opened up the world to anyone with a radio.

Lee De Forest. Born in Council Bluffs, Iowa, Lee De Forest was six when his minister father became the first president of Talladega College and the family moved to Alabama. The young De Forest worked part of the summer of 1893 at the World's Columbian Exposition in Chicago, an experience that inspired him to pursue an education in science and technology.

The following autumn, De Forest received a scholarship to Yale University. He earned a Ph.D. in physics, writing his dissertation on the reflection of radio waves. He then worked at several different electrical jobs, though none satisfied him. Three years after graduating, De Forest convinced several stock promoters to establish a wireless telegraphy company and soon demonstrated equipment that could transmit a dot-dash signal at least six miles. Eager to gain independence from foreign wireless companies, the federal government gave De Forest several contracts.

De Forest worked to improve vacuum-tube diodes. He had a display at the 1904 world's fair in St. Louis that included a 300-foot-tall antenna, the tallest structure at the fair. Partly because of his ability to send wireless messages to Chicago, De Forest received the fair's grand prize in wireless telegraphy.

Like most electrical investigators, De Forest hoped to find a more sensitive detector. He wanted to invent a device or circuit that would produce clear reception of long-distance transmissions. After much trial and error, De Forest placed a zig-zag-shaped piece of nickel wire, which he called a *grid,* between the anode and cathode of a diode. With proper circuitry, the three-element tube amplified Morse code far better than any diode. Calling his tube the *audion,* or *triode,* De Forest took out a patent on it in 1907. This was the prototype for billions of radio tubes that followed.

De Forest married four times. His fourth and happiest marriage in 1930 was to silent-movie star Marie Mosquini, who shared his enjoyment of music, poetry, and camping. De Forest received over 300 patents, his last granted at 83, just four years before his death.

The three-element triode paved the way for the four-element tetrode, the five-element pentode, and many other sophisticated amplification tubes. All used large amounts of power, partly because of their glowing filaments. As transmitting demands became more and more challenging, circuits became larger and more complex, and they required more electrical power. Few electronic devices were small and lightweight enough to be truly portable. That changed when William Shockley (1910-1989) and two other researchers at Bell Labs invented the transistor in 1947.

William Shockley. Born in London, England, William Shockley was the son of a mining engineer father and a mineral surveyor mother. Though Americans, they were in Britain on a company assignment at the time of Shockley's birth. He was home schooled back in Palo Alto, California, and became interested in physics through a neighbor who taught at Stanford University. Shockley received degrees from the California Institute of Technology and the Massachusetts Institute of Technology (MIT). After earning a Ph.D. from MIT in 1936, he joined the staff of Bell Telephone Laboratories in Murray Hill, New Jersey.

At Bell Labs, Shockley eventually headed a team that included John Bardeen (1908-1991) and Walter Brattain (1902-1987). Shockley's team worked with diode crystals that did not require electrical power, as the Fleming valve did. The two most common crystals, silicon and germanium, were called *semiconductors.* They conducted current better than insulators but not as well as conductors. Bell Lab scientists became interested in the crystals in the 1930s. They hoped to use semiconductors with ultra-high-frequency radio wave communication.

Shockley's team prepared very pure crystal samples with small amounts of precisely measured impurities, such as phosphorus

This device, which looks like a jumble of wax and wires, is a replica of the first transistor operated at Bell Labs in 1947.

The 1954 Regency pocket radio was the first transistorized consumer product. It used four germanium transistors and a 22.5 V battery.

and iridium. The use of phosphorus resulted in an n-type crystal, which meant that the crystal had negative characteristics. The use of iridium resulted in a p-type crystal, which meant it had positive characteristics.

Shockley's team did countless experiments with different arrangements before they arrived at the right combination in December 1947. They started with a tiny piece of n-type germanium with one wire connected to its bottom. They made two gold contacts, less than 0.002 inch apart, at the top. Voltages, signals, loads, and instruments were connected to the small device. The group had created a solid-state device that amplified their voices by a factor of 40.

The researchers called their invention a *transistor* from the concept of "transferring current across a resistor." J. R. Pierce, a coworker who became director of electronic research in 1952, suggested the term. Shockley, Brattain, and Bardeen had made a *point-contact transistor*. Some later types were *junction transistors,* which sandwiched n- and p-type materials in three layers with three wire leads. Transistors were small and did not have the large voltage and power requirements of vacuum tubes. For their discovery, the team members shared the 1956 Nobel Prize in physics.

Shockley was a competitive individual who wanted to capitalize on his co-invention. He returned to Palo Alto, where his mother still lived, in 1956. Up to that time, nearly all transistors used germanium because it was easier to prepare in pure form. But silicon was plentiful and less expensive. Shockley found financing and established the Shockley Semiconductor Laboratory to produce a practical silicon transistor. He was the person who started what came to be called "Silicon Valley."

Shockley was hard to work with. Several employees who left Shockley set up companies of their own in the area south of San Francisco. Nearly all the companies in the Silicon Valley trace their technical lineage to the Shockley Semiconductor Laboratory.

Shockley left the electronics industry in 1963 to accept an appointment at Stanford University. He became a controversial figure in the later 1960s for expressing unorthodox and controversial views on genetics. He died at 79.

It is often a long jump from a successful laboratory experiment to a production unit. But the great potential of transistorized products encouraged their rapid development. Transistors made by the Western Electric Company were available as early as 1951. Three years later, the pocket-sized Regency TR-1 radio went on sale for $49.95. Made by Texas Instruments and Regency Electronics, it was the first transistorized consumer product.

Integrated Circuits. Transistors were far smaller than the vacuum tubes they replaced. They encouraged the creation of more and more sophisticated electronic machines, such as computers. The computer's complexity required the use of more transistors, along with the resistors and capacitors that were part of the circuits. The next step in the continuing electronic revolution involved the *integrated circuit (IC),* which combined all components and wiring on a single crystal made of semiconductor material. Commonly called a *microchip,* or simply *chip,* it was developed independently in 1958 by Jack Kilby (1923-) at Texas Instruments and Robert Noyce (1927-1990) at Fairchild Semiconductor Corporation.

Jack Kilby. Jack Kilby was born in Jefferson City, Missouri. His father became president of a Kansas power company and

Kilby grew up in Great Bend, Kansas. After graduating from the University of Illinois with a degree in electrical engineering, he went to work for Centralab, a Milwaukee electronics manufacturer. In 1952, the company sent Kilby to Bell Labs to participate in workshops on germanium transistors. Centralab wanted to use the transistors in hearing aids.

Kilby moved to Texas Instruments (TI) in the summer of 1958 to work on miniaturized components, which TI called *micro modules*. The day he started work, everyone in his area was on vacation. As a result, Kilby was free to look at alternatives to accepted TI techniques. While TI stacked components vertically, Kilby set to work designing methods for a horizontal layout. He wanted to combine all the electronics in one device. He diffused impurities into a silicon chip to create several transistors. He believed there was no need to wire anything together since all the connections would be made inside the chip. By the following winter, Kilby had a working model that was half the size of a paper clip.

Kilby and TI filed for a patent in February 1959. The next month, TI introduced an early IC module priced at $450. In 1960, a transistor occupied the space of a 5 mm cube. Four years later, ICs allowed for fitting 100 transistors in the same area.

Kilby worked with two other TI engineers to develop the first handheld electronic calculator. Kilby, Jerry D. Merryman, and James H. Van Tassel completed their prototype in 1967. TI introduced the first self-manufactured, four-function, handheld calculator in 1972. The TI-2500 DataMath cost $150. Dozens of other manufacturers began selling calculators at about the same time and the competition was fierce. In 1972, Hewlett-Packard came out with the first calculator to go beyond the four functions of adding, subtracting, multiplying, and dividing. The HP-35 scientific calculator cost $400 and at one point Hewlett-Packard had a waiting list that held up purchases for nearly six months. TI struck back in 1976 with its $25 TI-30. It proved the best-selling scientific calculator of the 1970s.

Kilby left TI in 1970 to work as an independent consultant and inventor. He held about 50 patents. He served as Distinguished Professor of Electrical Engineering at Texas A&M University from 1978 through 1985. In 2000, he won the Nobel Prize in physics for his part in the invention and development of the IC.

Robert Noyce. Born in Burlington, Iowa, Robert Noyce was the son of a minister. He grew up in Grinnell, Iowa. A friend who taught physics at Grinnell College encouraged Noyce to study mathematics and physics. He earned a Ph.D. from MIT in 1953, then joined the Philco Corporation in Philadelphia to work on the development of transistors. A few years later, he went to work at William Shockley's company near Palo Alto, California.

Noyce was one of the eight employees who left the company in 1957 over disagreements with Shockley. They co-founded the Fairchild Semiconductor Corporation, one of the early companies in Silicon Valley. Noyce worked on integrated circuits at the same time that Kilby did. While Kilby worked with germanium, Noyce based his experiments on silicon. The two men applied for patents within a short time of each other.

Following several years of legal battles, TI and Fairchild decided to cross-license their technologies. In 1969, the U.S. Court of Customs and Patent Appeals ruled in favor of Noyce's claim of priority. Nonetheless, both companies succeeded partly because they could supply components for the Apollo moon-landing program, which President John Kennedy had just announced. Such a venture would require advanced electronic devices to control sophisticated systems on the Saturn 5 rocket, the command

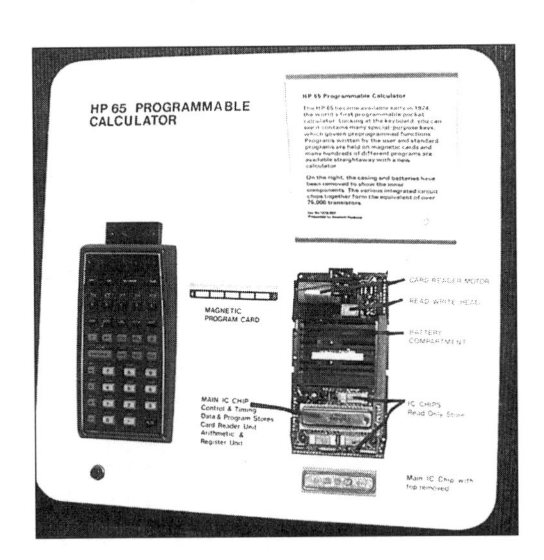

The Hewlett-Packard 1974 HP-65 was the first programmable calculator. Lines attached to the cutaway at the right indicate various internal parts. The one in the middle points to the "main IC chip" and the one at the lower right points to "IC chips."

module, and the lunar excursion module.

Noyce left Fairchild in 1968 to cofound the Intel Corporation with Gordon Moore (1929-), one of his Fairchild cofounders. The name Intel came from "integrated electronics." Intel made a complete microcomputer by putting all the logic into a single device called a *central processing unit*. The company produced the Intel 4004 in 1971. The first single-chip microprocessor, it sparked the growth of the personal computer industry. Noyce served as president of Intel until 1975 and then chairman of the board until 1979. He held 16 patents for semiconductor devices, production methods, and structures. He died at age 62.

References

"Driving the Dynamo" by John Bowditch, *Mechanical Engineering,* April 1989.

"Genesis of the Transistor" by Walter H. Brattain, *The Physics Teacher,* March 1968.

Benjamin Franklin by Thomas Fleming, Harper and Row, 1972.

Faraday Rediscovered, edited by Frank Gooding, American Institute of Physics, 1989.

The Real McCoy–African-American Invention and Innovation by Portia P. James, Smithsonian Institution Press, 1989.

Michael Faraday of the Royal Institution by Ronald King, Royal Institution Press, 1973.

The Life of Charles Proteus Steinmetz by Johnathan Norton Leonard, Doubleday Publishers, 1929.

Empire of the Air by Tom Lewis, Harper Collins Publishers, 1991.

"Seeking Redress for Nikola Tesla" by Eliot Marshall, *Science*, 30 October 1981.

"How the Computer Got Into Your Pocket" by Mike May, *American Heritage of Invention and Technology,* Spring 2000.

Those Inventive Americans, The National Geographic Society, 1971.

Measuring Invisibles, Weston Electrical Instrument Company, 1938.

"'St. George' Westinghouse" by Curt Wohleber, *American Heritage of Invention and Technology,* Winter 1997.

CHAPTER 3 Telegraph and Telephone

Introduction

People have communicated over distances for many centuries. The use of sunlight reflected from a polished surface and drumming provide two examples of early communication. Ancient Roman legionnaires along the 73-mile Hadrian's Wall, extending west from Newcastle, in northern England, used nighttime fires to communicate. Completed in 122 A.D., the wall had many forts, castles, and turrets, with signal stations located every few miles. Using a complicated coding system, commanders sent messages by lighting fires, which were then relayed all along the wall.

France's Claude Chappe devised a similar telegraph system using semaphores in the late 1700s. The term *semaphore* comes from *sema,* which means "sign," and *phorous,* which means "producing." Chappe located message transfer points high in the towers of tall buildings or on hilltops. Operators used wooden arms that resembled canoe paddles to relay messages up and down the line in stages. The method proved cumbersome and vulnerable to the weather. Nonetheless, it was the best available method and many countries adopted it. England installed a system of 15 stations between the port city of Portsmouth and London, a distance of about 60 miles. Residents of Massachusetts also used a semaphore system in the early 1800s. In many countries, the name "telegraph hill" still indicates places where the semaphore towers once stood.

The remnants of Hadrian's Wall can be seen from speeding British trains. Built by the ancient Romans, the wall is now mostly ruins and is used to separate farms. In its heyday, legionnaires communicated by lighting nighttime fires according to a secret code.

Samuel Morse provided a better system when he patented his electric telegraph in 1840. But he was not the first to succeed at electric telegraphy, and he did have serious competition from England's Charles Wheatstone. Wheatstone had the first practical system operational in 1837. By 1839, it connected Liverpool with Manchester. Morse's method won out, though, because of its simplicity and practicality.

Telegraphs used a fairly complicated code of dots and dashes to send messages. Although a skilled operator could typically send about 50 words per minute, the telegraph did not lend itself to use by people in their homes. But the telephone did.

Like the telegraph, the telephone involved a point-to-point wire-based commu-

Although no longer professionally associated with the telephone after the late 1880s, the popular Alexander Graham Bell was often on hand for photo opportunities. Here he is shown speaking into a telephone at the 1892 opening of the line between New York City and Chicago.

Claude Chappe of France developed the first uniform telegraph system in 1791. The nonelectric system used a semaphore signaling technique.

Pioneers of Electrical Communication by Rollo Appleyard, Macmillan, 1930

nication method. Alexander Graham Bell received his telephone patent in 1876. It is often described as the most valuable patent issued anywhere in the world. Later, many other people added improvements to the telephone. By 1900, more than 3,000 patents related to the telephone existed. Thomas Edison developed an improved mouthpiece and started the custom of using the word "hello" when answering the telephone. It was a more formal but previously little used version of "halloo." Bell had suggested the word "ahoy" as a telephone greeting, but it did not catch on with the public.

At the close of the 20th century, there were almost 250 million telephones in the United States. About a quarter of them were wireless cellular telephones. Other technologies also piggybacked onto the telephone infrastructure, including facsimile machines and Internet connections. Wired or wireless, the telephone certainly changed how people communicate. It altered the way people conducted business transactions, and it helped friends and relatives stay connected. Modern high-speed communication began with those 18th century semaphore arms, which were in their day considered to possess the ultimate in transmission speed. The method they used was referred to as *visual telegraphy*.

Telegraphy

Semaphore signaling was the first transmission method to be referred to as a *telegraph*. The ancient Greek word *tele* meant "distant" and *graphos* meant "writing." A visual telegrapher stationed on a tower used a telescope to decode a message sent by a signaler several miles away. The person receiving would then turn cranks located inside the tower to align wooden arms on the tower's top. The position of the wooden arms relayed the message to a receiver at the next tower in the line, and so on. The system was introduced in France in 1791.

Claude Chappe. Born in Brulon, France, Claude Chappe (1763-1805) had three brothers: Abraham, Ignace, and Pierre. Their father, a lawyer, supported the family very comfortably. All the brothers were educated in a seminary, and Claude originally planned to take on the religious life of a

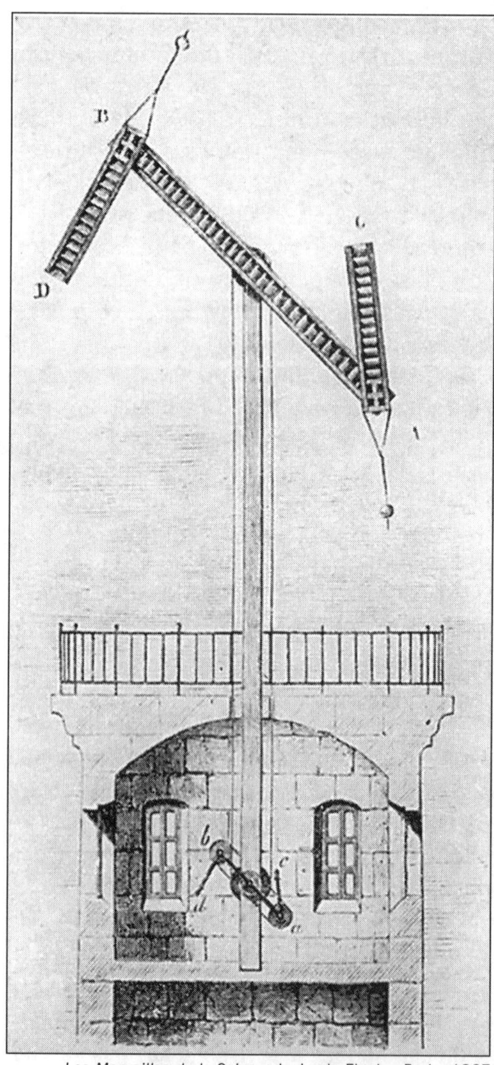

Claude Chappe's 1791 semaphore telegraph system used wooden arms at the end of a horizontal crosspiece. The telegrapher positioned the arms with cranks inside the tower.

Les Merveilles de la Science by Louis Figuier, Paris, 1867

cleric. He changed his mind around 1781, when he grew interested in technology. Chappe originally focused on electricity, carrying out investigations on electrically charged soap bubbles. Later, he and his brothers experimented with sending messages between two stations, a quarter mile apart. At their first public demonstration, in Brulon in March 1791, they sent messages over a distance of 15 km.

The Chappes finally devised a workable system during the 1790s that connected all of France's major cities. Their stations were stone towers that resembled windmills. A heavy wooden pole, with a rotatable horizontal crosspiece, rose about 10 feet in the air. The Chappes attached wooden semaphore arms to the ends of the crosspiece. An operator rotated cranks inside the tower, which controlled the orientation of the semaphore arms. A total of 196 positions could be interpreted at a distance. Many positions were coded to indicate a word or a phrase, and completed messages were called *bulletins telegraphiques*.

Placing lanterns at the ends of the signaling arms sometimes made nighttime transmissions possible. But fog and rain often defeated the system. For short messages in good weather, the visual telegraph worked quite well. Twenty-two semaphore signaling stations connected the 150 miles between Paris and Lille. Historian Louis Figuier wrote in the 1860s that talented telegraphers could send a short message in as little as two minutes. That time seems a bit optimistic, but it does suggest that Figuier felt impressed with the system's speed.

At its peak in 1852, the Chappe system in France had 19 branches and 556 stations, and it ran a total length of 4,000 km. But Chappe found that establishing the basic network took considerable organizational, political, and diplomatic skills. As with almost all successful inventions, others claimed to have had the idea first. That continual stress proved more than Chappe could tolerate, and he took his own life in 1805.

Residents of the lower Cape Cod region of the United States also used Chappe's method. Jonathan Grout installed a 65-mile system in 1800. It transmitted the news of incoming vessels between the island of Martha's Vineyard and Boston. A lack of subscribers ended the service in 1807, and the shortcomings of visual telegraphy encouraged people to look for other methods. The new technology of electricity offered promise. In Germany, Samuel Soemmerring (1755-1830) developed a telegraphic readout method based on converting water into hydrogen and oxygen. Andre Ampere (1775-1836) suggested using the swing of a galvanometer needle. But the first innovators to achieve success were Charles Wheatstone (1802-1875) and his partner William Cooke (1806-1879).

Charles Wheatstone invented and manufactured bellows-operated concertinas in the early 19th century. This privately owned one was made at Wheatstone's shop in London and is disassembled to show its reeds.

Charles Wheatstone and William Cooke. Charles Wheatstone's father, who worked in the retail music trade, moved his family to London when Charles was four. There, the young Wheatstone learned about music, sound transmission, and wave propagation. He received primary education at a private school, then apprenticed in 1816 to an uncle in the music business. He had no formal technical instruction.

When Wheatstone's uncle died in 1823, Wheatstone and his brother William took over the business. They specialized in making flutes and other air-moving instruments. Wheatstone invented and patented the concertina in 1829. The brothers had others produce the concertina parts, which they assembled and sold as completed instruments. An experimenter at heart, Wheatstone also built a small harp that he suspended from the ceiling with a thick wire. A piano on the floor above the harp was attached to the wire, which produced music from the harp when the piano was played. Wheatstone found that he could use that mechanical technique to send music and speech over considerable distances. But he concluded that electrical methods offered more hope than mechanical ones.

While Wheatstone investigated the field of electricity, he did not neglect his other interests. He measured the speed of electrons in a conductor and tried to slow down a spark. He examined light given off by burning metals, a practice now called *spectrum analysis*. Wheatstone also invented the *stereoscope* in 1832. Resembling a pair of binoculars, it allows the user to view two almost-identical images at the same time to produce a single image that appears three-dimensional. Wheatstone occasionally wrote technical papers and made presentations. In 1834, he was appointed Professor of Experimental Philosophy at King's College in London. He married Emma West in 1847 and they had five children.

William Cooke was the son of a surgeon and professor of anatomy at Durham University. Like Wheatstone, he had no formal technical education. While convalescing from an illness, Cooke traveled to Europe, where he learned about the idea of instantaneous, long-distance communication. Early on, he had little success with the concept and was advised to contact Wheatstone. Michael Faraday introduced the two men, setting up an ideal match. The combination of Cooke's strong business drive with Wheatstone's scientific skills paralleled the successful Matthew Boulton-James Watt steam engine partnership.

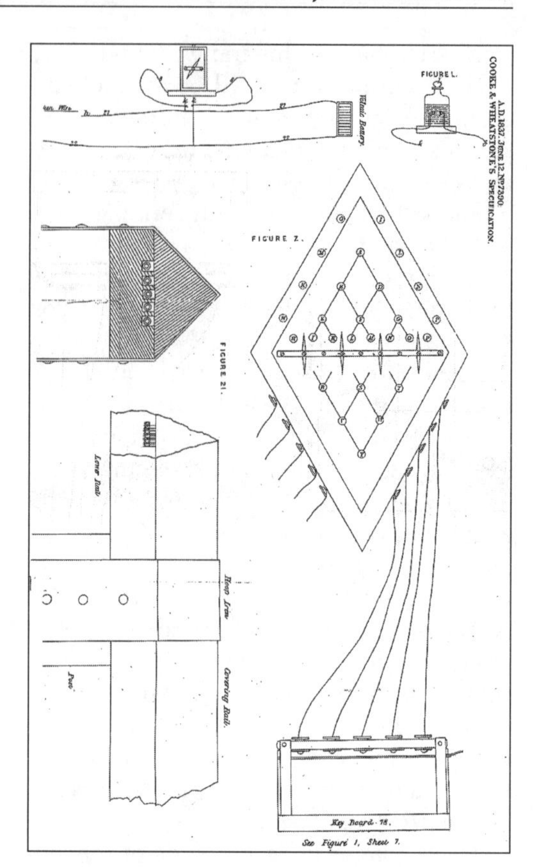

British patents often include drawings that appear on several large sheets of paper. This drawing of a section from Wheatstone and Cooke's 1837 telegraph patent was pieced together from two sheets.

Wheatstone and Cooke pooled their ideas to develop an indicating telegraph. Instead of using dots and dashes, it used five needles that pointed out specific letters on a diamond-shaped lattice. Wheatstone and Cook received a joint patent in 1837 for the world's first practical telegraph. They constructed their first long-distance line over a two-mile length in the northwest area of London. Next they put up a 13-mile line along the Great Western Railway running between London's Paddington Station and West Drayton. By the early 1850s, Britain had over 4,000 miles of telegraph lines.

The Wheatstone-Cooke system used six wires, one for each needle and a return wire. Sending a letter, such as "P," caused two needles to deflect slightly. They would point in the direction of two lines, at the intersection of which would lie the letter "P." Wheatstone and Cooke established the Electric Telegraph Company. Their system linked the port city of Liverpool with the manufacturing city of Manchester in 1839. It was the world's first public telegraph service. Opera-

This five-needle 1837 Wheatstone and Cooke telegraph was a direct-reading instrument. Two needles deflected and their combined direction pointed out a specific letter. In this photo, the left and right needles point to the letter A at the top. The telegraph used a 20-letter alphabet.

tors, who required little training, could send about 22 words per minute.

Railways had a particular interest in rapid communication. The 1830 Liverpool and Manchester Railway was the first major rail line in the world. Early railway lines used only a single set of tracks, and operators needed the ability to communicate about late departures or damaged equipment. So the British railways made extensive use of the Wheatstone-Cooke system.

Far from working well together, though, Wheatstone and Cooke often quarreled. Each saw himself as the sole inventor of the electric telegraph. The partnership broke up in 1845. Each man had made a fortune through their telegraph, but Cooke lost his in later unsuccessful ventures. He died at 73 in Farnham, a few miles southwest of London.

Wheatstone is commonly identified with the Wheatstone Bridge for measuring resistance, a circuit he did not develop and one for which he never claimed credit. It was actually devised in 1833 by his close friend Samuel Hunter Christie (1784-1865). Wheatstone's name became associated with the bridge circuit because he often used it in his experiments.

Wheatstone worked nonstop. His other successes included the invention of the variable resistor in 1843. He improved the dc generator by making one that delivered a more constant current. In 1844, he became the first person to experiment with underwater telegraph cables. He invented the chronoscope, an instrument that measured small time intervals. He received at least 34 awards of distinction. Wheatstone died while on a business trip to Paris. Queen Victoria knighted both Cooke and Wheatstone for their accomplishments, but history has tended to side with Wheatstone as the major innovator in the partnership.

The indicating electric telegraph remained popular for several years. But the delicate sending and receiving equipment required very careful use. Also, the system originally required six wires, though that dropped to two wires following the use of a code. However, Wheatstone and Cooke's system soon lost out to a simpler and more reliable system developed in America by Samuel Morse (1791-1872), who also introduced the Morse code.

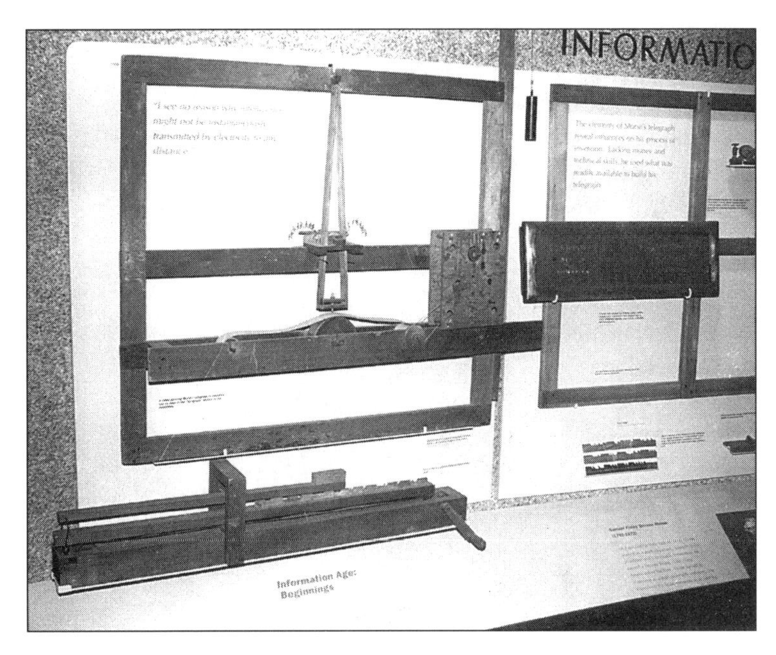

Samuel Morse provided this model to accompany his patent application, which was granted in 1840. The wooden device at the bottom was Morse's original transmitter. The notches cut into the strip of wood represented letters.

Samuel Morse and Alfred Vail. The son of a minister, Samuel Morse was born near Boston. Both of his parents appreciated the value of education. Morse attended local schools and an academy for gifted youngsters. He majored in art at Yale College but also took elective classes in electricity, chemistry, and mathematics. After graduation, his parents financed a trip to London so that he could study under the leading painters of the time. Morse had real artistic talent. Yale University and the New York Public Library currently own a number of his paintings. Morse married Lucretia Walker and they had three children.

Morse spent some time back in America, then in the late 1820s returned to Europe. After meeting Louis Daguerre (1787-1851) and learning the daguerreotype process, Morse again headed back to America. During the return sailing, he discussed the subject of electricity with a fellow passenger. Morse's new acquaintance mentioned electricity's potential for sending information over any length of wire instantaneously. The idea fascinated Morse, who set to sketching a device that he could use to send electrical messages.

Back in America, Morse taught painting and sculpture at the newly opened New York University. He was America's first professor of art. He also opened one of the first photographic portrait studios, where he made daguerreotypes, and worked on the telegraph in his spare time. Morse spent

A close up of the moving parts on a replica of Samuel Morse's 1837 printing telegraph. A strip of paper passed under a pen that was deflected by an electromagnet. The pen made a series of peaks and plateaus that the telegrapher interpreted as dots and dashes.

seven years trying different telegraph designs, selling partnerships and some of his paintings to pay the bills.

The device Morse patented in 1840 did not resemble the standard telegraph receiver common a few years later. The original device was, in essence, a printing instrument. Morse used a small wooden frame to support a pendulum that had a pencil attached to its end. One cross piece near the bottom had a roll of paper tape and a clockwork mechanism that pulled the paper under the pendulum. Morse used a series of dots and dashes as his transmission code. Rapidly closing the transmitter key caused an electromagnet to move the pendulum quickly, producing a dot. Closing the key more slowly made a dash. It was a cumbersome system, but it worked. By 1845, trained telegraphers listened to the dot-dash pattern produced by an electromagnet in a sounding box and wrote messages down directly.

Needless to say, not all successful inventors worked alone. Chappe had his brothers, Wheatstone had Cooke, and Morse had Alfred Vail (1807-1859). Vail was born in Morristown, New Jersey, into a well-to-do family. His father owned the Speedwell Iron Works. Vail graduated from the City University of New York in 1836, but ill health kept him from his plan of joining the ministry. He became fascinated with the telegraph after attending an exhibition given by Morse in 1837. Because of financial backing promised by Vail's father, Morse took Vail in as a junior partner.

Vail had a patent model built at his own expense and constructed a three-mile-long experimental telegraph line around his father's factory grounds. He agreed to let Morse demonstrate the telegraph by himself in Washington, D.C., and to underwrite other expenses associated with obtaining overseas patents. With Vail's father providing the necessary funds, Morse filed for his first patent in 1838. He received it in June 1840 as Patent Number 1,647. Vail's name does not appear on any of the patent's seven pages.

Vail may actually have devised the Morse code. He certainly evaluated the relative frequency with which letters appear in the English language. He also suggested that the letters that appeared most frequently should have the shortest code representation. For example, a dot represents "E," the most common letter. A dash represents "T" and a dot plus a dash represents "A." The less-common letters use the longer dot-dash combinations. Historians disagree on Vail's role in developing the Morse code.

Vail also made improvements to some of Morse's hardware designs. These included an improved electromagnet and vibrating circuit breaker. He also made working drawings for an ammeter used in the circuit. However, the concept of the electromagnet telegraph clearly belonged to Morse, and Vail never seemed concerned by his lack of recognition. He remained with Morse until about 1848, when he lost interest in telegraphy. Vail died at 52 in Morristown.

The United States Congress approved a $30,000 grant for a 41-mile trial telegraph line between Washington and Baltimore.

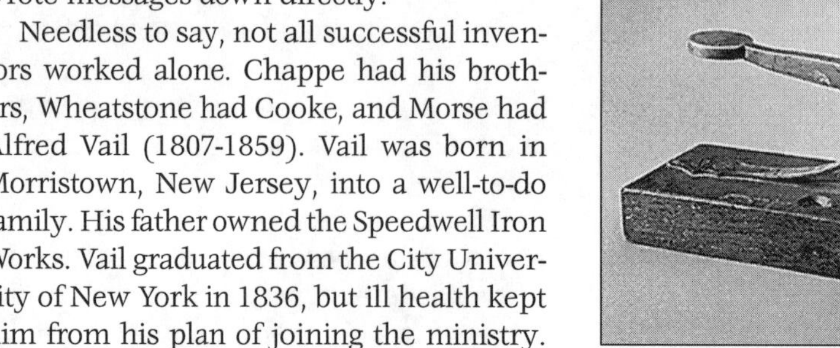

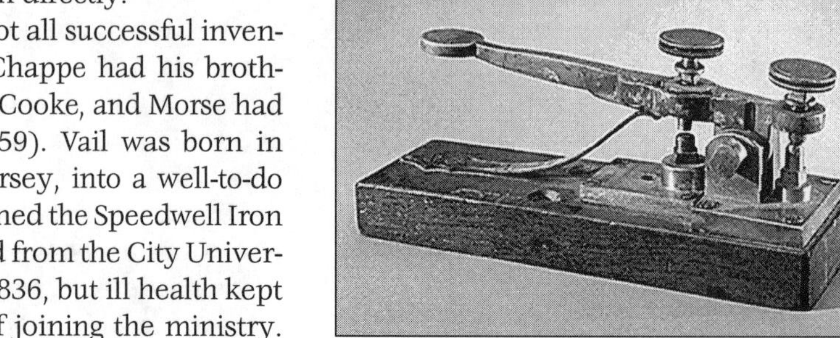

Smithsonian Institution

Samuel Morse used this telegraph key to tap out his first official inter-city message. The message went from Washington, D.C., to Baltimore in 1844. The key was an improved version of Morse's 1837 patented design.

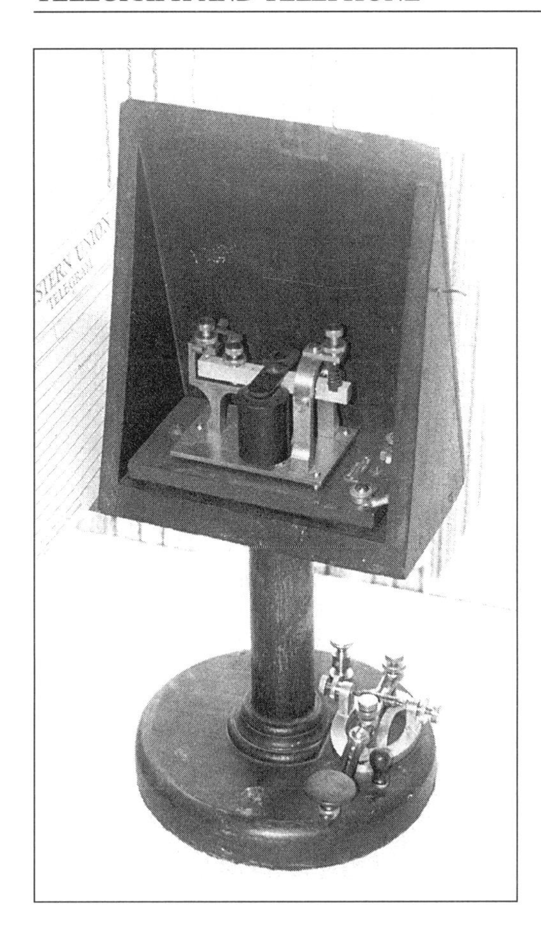

This 1920s-era telegraph sounding box acts as an amplifier and speaker for the receiver. The transmitting key on the pedestal base was made around 1900.

Morse accepted a suggestion to use a single wire strung on glass insulators casually mounted on trees and poles. Using the earth as a ground eliminated the need for a two-wire system. The telegraph's terminal points were the Supreme Court chambers and the Baltimore and Ohio Railroad depot. On May 24, 1844, from Washington, Morse tapped out "What hath God wrought?" It took him one minute to transmit the four words to Vail in Baltimore.

Just four years later, with the exception of Florida, every state east of the Mississippi River had telegraph service. Telegraph lines connected the American coasts in 1861. Morse lived to see all the early major developments in telegraphy. He died at 80 in New York City.

In the field of technology, everything has a beginning and an end. In 1998, satellite receivers replaced the use of the Morse code for the communication involved in international sea travel. At that point, all ships had to carry automatic distress beacons.

Before wireless transmission, telegraphy offered the only way to communicate in real time over a great distance. Overland transmission required stringing wire overland from pole to pole, and problems multiplied many times over when water was involved, especially the Atlantic Ocean. Undaunted, Cyrus Field (1819-1892) connected North America to Europe in 1866 with a permanent undersea telegraph cable. It took seven attempts and few people thought Field had any chance of succeeding.

Cyrus Field. Born in Stockbridge, Massachusetts, Cyrus Field decided at the age of 15 not to attend college like his older siblings. He went to work at a wholesale paper dealership in New York City and became a junior partner at 21. Field almost single-handedly worked the organization out of bankruptcy. But long hours of work left him exhausted and in poor health. With a personal fortune of $250,000, he retired from the business at 33.

Then, Field met a Briton who was working on a telegraph line in Canada. Their conversations made Field begin to consider the possibility of laying a connecting cable across the Atlantic Ocean. He discussed the project with Samuel Morse, who had no doubt that a signal could be transmitted across the 2,300-mile underwater distance. The shortest distance for a transatlantic cable would require that it come ashore in Newfoundland.

The underwater cable had a conductor with seven strands of copper wire. The insulated conductor was strengthened with spirally wrapped iron wires. Hemp and tar sealed the entire cable, which weighed 2,000 pounds per mile. Huge ships were needed to carry it. America and Britain each provided a ship with special cable-handling equipment and trained personnel. In Ireland, during the summer of 1857, workers loaded each ship with about 1,300 miles of cable. The plan called for the American ship to begin laying cable and, halfway to Newfoundland, splice it to the cable on the British ship.

Complex machinery let out the cable as the ships slowly sailed west at 4 mph. But after only 335 miles, the brakes grabbed too firmly and the cable broke. It fell, unretrievable, into the ocean. Field used his persuasive skills and organizational ability to arrange four more attempts. The fourth succeeded temporarily. On August 16, 1858, Queen Victoria sent the first official transatlantic telegraph message to President James Buchanan. Her 90-word greeting took over 16 hours to send and verify. At a maximum transmission rate of only four words per minute, transatlantic telegraphy proved much slower than land telegraphy.

Still, about 400 messages were transmitted before the cable broke on September 1. The economic depression of 1860 and the Civil War prevented further efforts. But Field had tremendous personal drive and never gave up on the project. He worked on a stronger cable design, increasing its weight to over 3,300 pounds per mile. Before the Civil War ended, Field engaged the services of the largest ship in the world, a 692-foot-long British steamship. Isambard Brunel's *Great Eastern* had five times more capacity than the next-largest ship at the time. Brunel (1806-1859) had designed it to transport 4,000 passengers to the East Indies without having to refuel. Its huge size made it the ideal vessel to carry all the underwater cable needed to span the ocean.

The *Great Eastern* began laying cable from Ireland in 1865. About 600 miles from completion, the cable broke and fell to the ocean floor. Grappling equipment could not lift the broken end of the new heavier cable. But even this sixth failure did not deter Field. The next year he made another attempt, which proved successful. Britain's great 19th century scientist Lord Kelvin (1824-1907) was on the ship as an interested observer when the cable was permanently completed on July 27 between Valentia, Ireland, and Hearts Content, Newfoundland. Within days, people stood in line to send transatlantic messages at $5 to $10 per word.

Lord Kelvin's small 1860s mirror galvanometer made the transatlantic telegraph cable practical. This one was used at the North American end.

However, telegraphers soon had problems receiving the weak electrical telegraph messages from the lengthy underwater cable. They found messages very hard to decipher and had to request verification time and time again. The person in charge of signal transmissions felt that the cable would operate better at high voltage. High voltage, during an unnecessary test, damaged the cable, but did not put it out of service. After a number of arguments, both legal and technical, Lord Kelvin installed a low-voltage readout system that used a highly sensitive mirror galvanometer he had invented. About the size and shape of an ordinary flashlight battery, it had a tiny magnet and mirror hanging by a silk thread inside a coil of fine copper wire. Weak transatlantic signals arriving at the coil caused the mirror to rotate slightly. A beam of light reflected by the mirror multiplied the effect. Motion of the light beam to the left or right indicated Morse code dots and dashes. Kelvin's mirror galvanometer rescued a huge financial investment and Queen Victoria knighted him for his contribution in 1866. The mirror galvanometer was one of Kelvin's 70 patents.

For his part, Field received innumerable awards and other forms of recognition. He went on to assist with cable-laying in other countries. He also helped with the elevated railroads in New York City, devoting much time to their completion. Parents of seven children, Field and his wife, Mary, celebrated their golden anniversary in 1890. Field died two years later at 72.

Telegraph traffic increased more rapidly than most people expected. Nearly 400 million telegrams per year were sent in Britain at the end of the 19th century. An early problem was that only one message could be sent at a time. Long waits to send or receive a message minimized the advantages of the new instant communication. Many people worked on ways to send several messages at once along the same wire. The practice was called *multiplexing* and various inventors had some measure of success. Jean Baudot (1845-1903) developed the most in-

Alexander Graham Bell was born in 1847 in this townhouse on Charlotte Street in Edinburgh, Scotland.

novative method, *time division multiplexing*, in France in 1874.

Jean Baudot. Born into a farming family near Magneux, Jean Baudot had no formal education beyond primary school and he began his working life as a farmer. Then, seeking adventure, he joined the French Telegraph Service at age 24. Baudot rose through the ranks, served in the Franco-Prussian War, and became an officer stationed in Paris.

Higher-ranking officers encouraged Baudot to develop a system for time multiplexing several telegraph messages. Baudot developed a very ingenious open-ended method. His experimental device used synchronized clockwork-driven rotating switches at the transmitter and receiver. Each switch had five segments, which permitted sending five messages over the same wire at almost the same time. Baudot patented his invention in 1874 and the French Telegraph Service officially adopted the system in 1877. His invention allowed the almost simultaneous transmission of as many messages as there were commutator segments.

Baudot's patent also covered the idea of a 5-bit code that had equal "on" and "off" intervals. Each letter is represented by a five-unit combination of current-on or current-off signals of equal duration. Using five bits allows having 32 different characters. More efficient than Morse code, the Baudot code was widely adopted for international telegraphy in the 20th century. In the 1970s, it was superseded by 7-bit and 8-bit codes. The modern unit of signaling speed is called the *baud* in Baudot's honor.

Telegraph Codes. Telegraphy was a revolutionary way of communicating in the 19th century. However, its users had to exercise economy in usage because transmission fees commonly ran as high as $1 per word. This led to the code phrases and clipped verbiage in the 1880s referred to as *telegraphese*. Open codes were published in books like *Bloomer's Commercial Cryptograph* and *A, B, C Telegraph Codes*. The codes helped minimize the use of extraneous words and misreading of the sender's intent.

Salespeople for machine tool manufacturer Joseph Brown (1810-1876) transmitted "affray" to mean "What is the price of?" The code word "abbott" referred to the company's universal milling machine. Thus, an "affray abbott" message meant, "What is the price of the universal milling machine?" The sales staff at Edward Weston's (1850-1936) electrical meter company used "reptatus" for the company's 150/300 volt dc voltmeter. The code word "requital" referred to the 600 volt dc voltmeter.

Telegrams were brief and impersonal. Not so the telephone. It let people speak directly to each other, responding immediately to questions. Like face-to-face discussions, the telephone conversations brought more human personality to communication by wire. Telephone use developed an etiquette resembling that of a written letter. There was the initial "hello" and the closing "good bye." The tone of a speaker's voice paralleled the individuality of handwriting.

Telephone

The word "telephone" immediately brings to mind Alexander Graham Bell (1847-1922). Although Bell produced the first successful telephonic device in 1876, others worked earlier on related projects. The principle of the telephone sounds deceptively simple.

- Human speech causes a reed or diaphragm to vibrate in a transmitter.

A replica of Alexander Graham Bell's 1876 telephone patent model.

• The vibration effect is electrically transmitted down a wire.

• A receiver vibrates in tune with the original speech, and the listener hears a human voice.

Charles Boursel (1829-1912) in France outlined a method in 1854, although he never created the hardware for it. Germany's Philip Reis (1834-1874) constructed a device in 1861, but it transmitted only tones, not modulated speech. America's Royal E. House patented a method in 1868 but did not realize its potential to transmit speech. Also in America, Elisha Gray produced a potentially serviceable telephone at about the same time as Bell. But Gray had more interest in telegraphy, and he missed his opportunity.

Alexander Graham Bell. Born in a comfortable three-story townhouse in the bustling city of Edinburgh, Scotland, Alexander Graham Bell was the son of a man who taught hearing-impaired people to speak. He also wrote several textbooks on the topic. Bell's mother was a musician and painter who instilled in him an appreciation for sound and music. During his early years, Bell's mother taught him at home. He eventually decided to follow the career path taken by his father.

After studying at a Scottish academy and Edinburgh University, Bell went to work with his father in London in 1867. Disaster soon uprooted the family. Bell's younger brother died of tuberculosis in 1867, then his older brother died in 1870. Sacrificing a well-established career, Bell's father moved his family to Ontario, Canada, which he thought would provide a healthier climate for his surviving child.

A principal at a Boston school for the hearing impaired asked Bell's father to instruct her teachers in the use of his methods. Going in his father's place, Alexander spent three months at the school. He later visited other schools in the area and the demand for his services increased. Bell went on to open his own school for teachers. Through his work, he met a young deaf woman, Mabel Hubbard, whom he married in 1877. He became an American citizen in 1882.

As a part of his work with the deaf, Bell began developing an electric speaking telegraph in 1873. Finding that he lacked the skill to make the necessary parts for his experiments, he hired the young Thomas Watson (1854-1934) as an assistant in 1874. The two men developed a close friendship and Watson eventually received a share in the telephone patent as payment for his work.

Bell's patent application described a transmitter that had a short platinum wire attached directly to a diaphragm. As speech caused the diaphragm to vibrate, the wire moved up and down in a weak acid solution. When the wire went deeper, the resistance decreased. When the wire rose, resistance increased. Since the sound controlled the current though the wire, it was possible to send and receive human speech using this technique. Unlike the on/off characteristics of telegraphy, a telephone used continuous current whose intensity varied with sound.

This 1882 wall-mounted unit shows one stage in the evolution of the telephone's design. The battery-operated unit was made by the (Werner) Siemens and (Johann) Halske Company in Berlin.

Library of Congress

The *Detroit News* published this photograph sometime between 1915 and 1925, long after Bell's company made significant ergonomic improvements to the telephone. The man is holding a model mounted on wood, perhaps to suggest the relative ease of using more modern phones.

On March 10, 1876, Bell's telephone transmitted the first intelligible sentence. Reacting to a minor accident involving spilled acid, Bell called out, "Mr. Watson, come here. I want to see you." The mouthpiece of their experimental telephone picked up the message. Watson, in another room, heard the sentences perfectly.

Bell set up a display for his new telephone at the 1876 Centennial Exposition in Philadelphia. It failed to attract much attention until the Brazilian emperor, Dom Pedro II, requested a demonstration. Listening at the receiver, Dom Pedro exclaimed, "It talks!" Newspapers reported the event on the following day, which introduced the public to the telephone. The original Bell Telephone Company came into existence in 1877. A year later, 3,000 telephones were in use and the first telephone exchange opened in New Haven, Connecticut, with 21 subscribers. Bell was rich and famous before reaching the age of 30.

For his part, Watson had a very interesting and satisfying life after he left Bell in 1881. Since he didn't have to worry about money, he could take work that interested him. Finding himself unsuccessful at farming, Watson went on to open a successful machine shop. In the 1890s, he established a company in Boston to build naval vessels. The company's payroll grew to 4,000.

Watson and his wife studied geology together at the Massachusetts Institute of Technology and had a fossil named after them. They went to California and Alaska to look for gold. In 1910, Watson found work as a minor actor in a Shakespearean acting company. He also dramatized novels for the actors to perform. He spent his final years in Boston doing theater, dramatic readings, and lectures on geology and the telephone. Watson likely found his uneven life quite stimulating.

Bell's technical work in telephony ended in the 1880s. He cheerfully admitted that keeping up with its rapidly advancing technology did not appeal to him. The first transcontinental telephone line was established in 1914 between New York City and San Francisco. During opening ceremonies, Bell repeated his famous line to Thomas Watson in California, "Mr. Watson, come here. I want to see you." Watson humorously answered that he would be glad to come, though it would take a week. Bell died at 75 and was buried near his summer home in Nova Scotia.

During his lifetime, many claimants came forward to contest Bell's patent. He was involved in about 600 court cases, but the courts upheld all of his claims. Some had been filed on behalf of Elisha Gray (1835-1901).

Elisha Gray. Born into a farming family in Barnesville, Ohio, Elisha Gray heard of the first telegraph line at the age of 10. He then tried to make his own, and he often tinkered with homemade batteries and electromagnets. After his father died in 1847, Gray worked his way through high school and two years at Oberlin College as a car-

A rest area off I-80 near the Utah-Nevada border provides an overlook of the Bonneville Salt Flats. The plaque states, in part, "On 17 June 1914, the first transcontinental telephone line was completed near this point . . . making the last splices in the wires that joined east and west with voice communication for the first time."

Alexander Bain published this image of his fax machine in 1850. The rope provides the energy to rotate the drum at the right, which carries the image to be sent by telegraph wires.

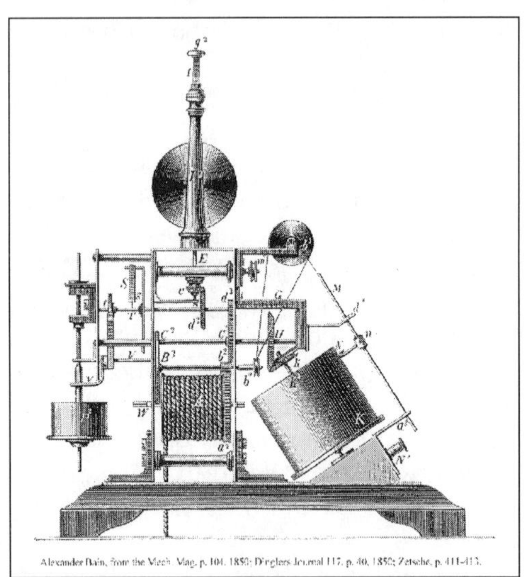

Mechanic's Magazine, 1850

penter. He studied electrical engineering. Gray married Delia Shepard in 1865 and they had one child.

Gray's first patent came in 1867 for a new telegraphic relay. To manufacture the relay, he established the Western Electric Company in Chicago. The company also made instruments for the Western Union Telegraph Company, the largest telegraph organization in the world. Gray's company prospered and he sold his shares in 1875 so that he could devote himself to full-time invention, primarily in telegraphy. He also remained a consultant to Western Union.

Like Bell and others, Gray had an interest in transmitting voice, but he spent most of his time with telegraphic devices. He had developed a metal diaphragm to transmit tones but saw little commercial value in the telephone. Gray heard of Bell's work in 1875 but did little to accelerate his efforts with voice transmission. Though opportunity knocked at his door, Gray ignored it.

Bell applied for his telephone patent on February 14, 1876. As luck would have it, Gray filed a formal caveat with the Patent Office a few hours later that day, stating that he intended to invent a telephone based on certain principles. As an American citizen, Gray could file a caveat. Still a British citizen, Bell lacked that right. So he applied for a full patent on an invention that he had already experimented with.

When the immense importance of the Bell patent dawned on Western Union, they contested it in court on Gray's behalf. The legal battle inspired charges that Patent Office employees had cheated Gray. But Bell had well documented his timing. After hundreds of pages of testimony, Western Union gave up its case as hopeless in 1879. In hundreds of court cases, other defendants attempted to revive Gray's claim in order to attack the validity of Bell's telephone patent. Every court case, including one argued before the United States Supreme Court, upheld Bell's patent.

Gray went on inventing items for electrical communication, eventually accumulating about 70 patents. One was for a *telautograph*, which transmitted facsimiles of handwritten letters and documents. It has been estimated that Gray earned more money from his inventions than did Bell. But Gray never forgot his near miss with history. He discussed it in a book he finished the year before he died at 66.

Morse's telegraph and Bell's telephone were both text devices. They dealt with the transmission of words. Just three years after Morse received his patent, a Scottish inventor received a British patent for sending pulsed information over telegraph wires. Alexander Bain (1818-1903) had developed the elements of the modern fax machine.

Facsimile Machines. Born in a remote area in the far north of Scotland, Alexander Bain came from a farming family. As a youth, he apprenticed to a watchmaker. After hearing a public lecture on electricity in the town of Thurso, he became fascinated by its possibilities. After his apprenticeship, Bain went to London as a journeyman watchmaker, in part because he wanted to learn more about electricity.

The first public demonstrations of the fax machine were conducted at the 1851 Crystal Palace Exposition with a device similar to this one. It was built by Frederick Blakewell.

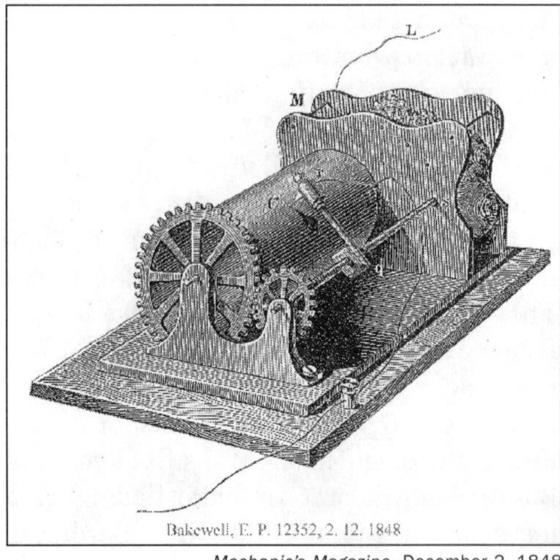

Mechanic's Magazine, December 2, 1848

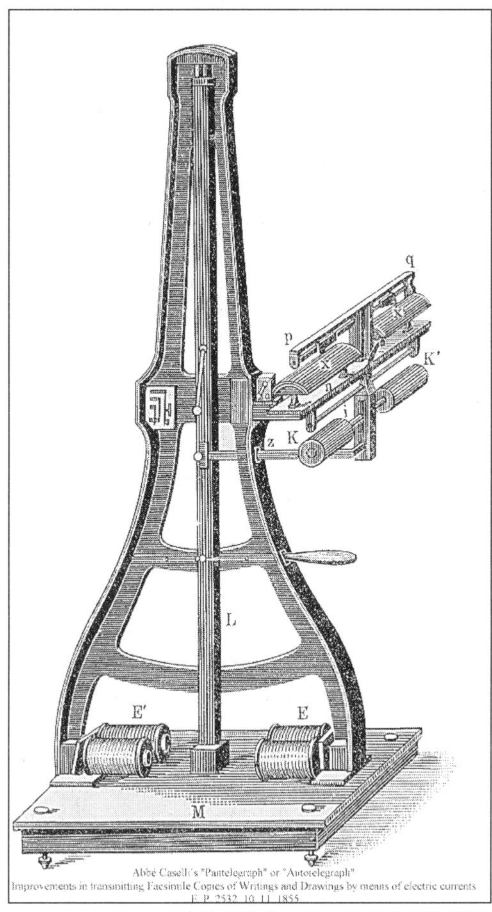

Mechanic's Magazine, November 10, 1855

In 1841, Bain received the first British patent for an electric clock. An oscillating internal part operated a mechanical switch that established an electromagnetic pulse. The regular pulsing allowed the clock to keep time. The elements of Bain's clock design also formed the basis for his concept of a facsimile (fax) machine.

Bain's numerous patents for electric telegraphy produced most of his income. His 1843 facsimile machine used timed pulses sent down a wire. In its simplest form, Bain envisioned two pendulums, in two separate cities, swinging in exact synchronization. At this early stage, Bain's invention, designed to send only text, functioned in the following way:

• The image to be scanned was composed of thin metal letters attached to a nonconductor like paper.

• The swinging pendulum at the transmitter had a small metal brush tip. The transmitter sent a pulse when the brush was in contact with the metal letters. No pulse was sent when it contacted the paper.

• The receiving pendulum also had a metal brush tip. It swung over paper treated with ammonium nitrate and potassium ferrocyanide.

• When an electric pulse was received, the current went through the chemically treated paper, darkening it.

• At both the transmitter and receiver, paper moved at a uniform rate under the swinging pendulums.

Bain receives credit for coming up with the idea of scanning an image so it can be arranged in small bits for transmission. But he never constructed a complete system. He could not synchronize the transmitter and receiver, which had critical importance. Other innovators would have to develop a workable system based on his patent designs. Bain's job title during his journeyman service was "mechanic," and he published his patented designs in *Mechanic's Magazine.*

British physicist Frederick Blakewell was the first person to publicly demonstrate facsimile transmission. Blakewell gave facsimile presentations at the 1851 Crystal Palace Exposition in London, the first world's fair. One of his changes to Bain's design involved writing the image to be scanned in nonconducting varnish on tinfoil. Blakewell reversed the technique that Bain patented.

In 1865, Italy's Giovanni Caselli (1815-1891) introduced the first commercial fax service between Paris and Lyon. He called his device a *pantelegraph* from the ancient Greek word *pan,* which means "all." Caselli used synchronized pendulums that measured about eight feet in height. His service lasted until at least 1870.

But it took several decades before photographs could be sent by wire. Black-and-white text message resulted in pulses that were either "on" or "off." Photographs, however, had shades of gray, which did not lend themselves to the technology of the time. Facsimile machine designers had to wait for the development of practical photocells in the late 19th century.

The first person to optically scan and send photographs was Germany's Arthur Korn (1870-1945), who made his first public demonstrations in 1902. Korn's transmitter used a selenium photocell to scan a photograph wrapped on a cylinder. The entire process

The first commercial fax machine service used two eight-foot pendulums. Giovanni Caselli sent text messages between Lyon and Paris beginning in 1865.

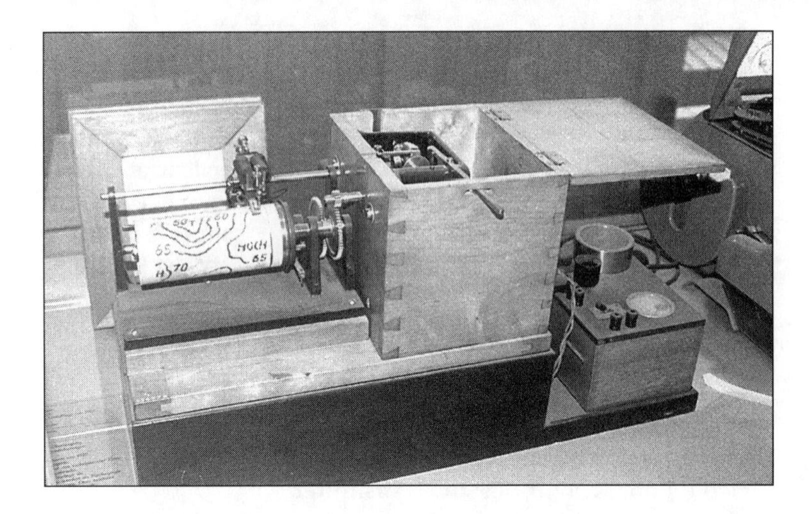

Germany's Max Dieckman made this 1928 facsimile machine. As the drum on the left rotated, it transmitted simple line drawings such as the weather map shown here.

took several hours per photograph. Nonetheless, Korn's invention went into service in 1906 transmitting newspaper photographs between Munich and Berlin. The word *wirephoto* soon came into use in newspaper jargon.

For many years, fax machines remained cumbersome and hard to operate. Then, in 1966, the Xerox Corporation introduced a small, practical office copier, the Magnafax Telecopier. This encouraged other manufacturers to get into the market and fax machine production boomed. Those different manufacturers established a variety of operability standards, but no international standards existed. This meant that American machines could not necessarily connect with European or Asian machines. In 1974, the International Telegraph and Telephone Consultive Committee issued its first worldwide standard known as Group 1 Fax. All Group 1 machines could connect with each other and print a one-page document in six minutes with a resolution of 100 lines/inch. The standards were upgraded over the years to allow for faster transmission times and improved resolution. It all began with Caselli's huge swinging pendulums in Paris and Lyon.

References

Pioneers of Electrical Communication by Rollo Appleyard, Macmillan Publishers, 1930.

The Communications Miracle by John Bray, Plenum Press, 1995.

History of the Atlantic Telegraph by Henry M. Field, Books for Libraries Press, first published 1866.

"What Hath God Wrought?" by Maury Klein, *American Heritage of Invention and Technology,* Spring 1993.

"No.1177: Thomas Watson" by John H. Lienhard, *Engines of Our Ingenuity,* National Public Radio, c. 1997.

French Inventions of the 18th Century by Shelby T. McCloy, University of Kentucky Press, 1952.

An Encyclopedia of the History of Technology edited by Ian McNeil, Routledge Publishers, 1990.

American Science and Invention by Mitchell Wilson, Bonanza Books, 1960.

MARCONI AND THE RADIO

CHAPTER 4 Radio and Television

Telegraphy was the 19th century's first means of rapid long-distance electrical communication. It went underwater in 1865 when a thin wire connected North America with Europe, as America's Cyrus Field completed the first permanent transatlantic telegraph cable. Germany's Werner Siemens, another communications pioneer, spent many years connecting London to Calcutta by telegraph over a distance of about 6,000 miles. He completed the line in 1870. Had wireless communication been available, Field and Siemens likely would not have tried such monumental undertakings.

Heinrich Hertz conducted laboratory experiments in Germany in 1886 that confirmed the theory of wireless signal transmission and reception. But it was left to Italian Guglielmo Marconi to send and receive the first wireless communication signal in 1895. Many others contributed to making radio a well-established technology during the 20th century.

Alexander Graham Bell's invention of the telephone in 1876 made other technologists consider the possibility of sending images by wire. Placed in the proper order, a series of dots can produce a high-quality image. The telephone was not a static technology. It delivered dynamic audio signals as people spoke them. Innovators made the leap to envisioning a system that would send video signals in real time. Even before hardware was available, the terms *telephonoscope* and *electric vision* came into use to describe what ultimately became *television*.

No single person can claim credit for inventing television. Dozens, from several countries, made significant contributions to making television a practical method of communication. John Logie Baird produced the first operational system, partly based on mechanical methods, in 1929. He received the world's first television license that year from the British Broadcasting Corporation. Russian-American Vladimir Zworykin demonstrated the first experimental all-electronic television system in 1924. But many other inventors also played a role.

Wireless pioneers worked in the field of signal transmission (telegraphy), voice communication (radio), and image transmission (television). The details of their work varied. But everyone's basic methodology went back to a remarkable series of experiments conducted in the late 19th century by Heinrich Hertz (1857-1894).

Heinrich Hertz. Born in Hamburg, Heinrich Hertz was the oldest of five children. His father, a lawyer and politician, provided the family with a comfortable living. His mother's relatives included physicians and religious leaders. After graduating from high school at the age of 18, Hertz went to work for a construction organization in Frankfurt. Later, he briefly attended a technical school and spent a year of military service with a railway regiment. Hertz then went on to

Guglielmo Marconi (left) looks at a telegraph tape with one of his assistants in a station that he built in 1899. Apparently quite conscious of his appearance, Marconi appears well dressed in almost all photos taken of him.

This replica shows how Hertz conducted his 1886 electromagnetic transmitter experiment. Electromagnetic waves came from sparks in the reflector at the left. They passed through the polarizing grating in the middle and were received inside the reflector at the right.

study physics at the University of Munich, transferring to the University of Berlin.

Hertz achieved his first technical triumph as a student in Berlin. A professional society offered a medal and cash award to the person who could solve a particularly difficult experimental problem that involved alternating currents. Hertz not only won the prize, but his solution later guided him toward the discovery of radio waves.

After receiving a doctorate degree in 1880, Hertz stayed on at the University of Berlin as a salaried assistant. He produced 15 publications and developed a noteworthy reputation. His technical skills touched on both theoretical and experimental aspects of electricity. In 1885, Hertz took a position at the Karlsruhe Technical University, where he would make his landmark discovery in radio transmission.

At Karlsruhe, Hertz met Elisabeth Doll, the daughter of a colleague. The couple married in 1886 and eventually had two daughters. Also in 1886, Hertz conducted the experiment that would make him world famous. He had noticed that Leyden jars, early capacitors, discharged with a spark across an air gap. A smaller spark across the terminals of a nearby Leyden jar would occasionally accompany the main spark. Hertz suspected that the primary spark caused the smaller spark. To test his idea, he constructed a high-voltage transmitter that included two metal spheres separated by 1 to 2 cm. He positioned the spheres at the focal point of a metal reflector and used a simple inductive circuit to produce high voltage that sparked across the spheres. The sparking transmitter created electromagnetic waves.

As his receiver, Hertz used a 70-cm-diameter loop of wire with smaller metal spheres at the ends, separated by about 3 mm. He tested the receiver in different parts of a darkened room, typically placing it 20 meters from the transmitter. Large high-voltage sparking at the transmitter induced weak sparks to jump the air gap at the receiver. Hertz became the first person to send and receive energy through space. Nothing like his experiment had ever been conducted.

Hertz was described as a quiet, pleasant person who rarely took credit for his accomplishments. In 1890, the British Royal Society awarded him its coveted Rumford Medal, a precursor to the Nobel Prize. Hertz told no one when he traveled to London to accept the award at formal ceremonies attended by the world's greatest scientists. He showed modesty to the point of never mentioning his discoveries, even when discussing their basic theory with students or colleagues.

Hertz was the first person to blend electricity into electronics, and in doing so he established the foundation for all wireless communication. But he had no premonition of the ultimate value of his work. When asked if he thought that electromagnetic energy could be used for wireless telegra-

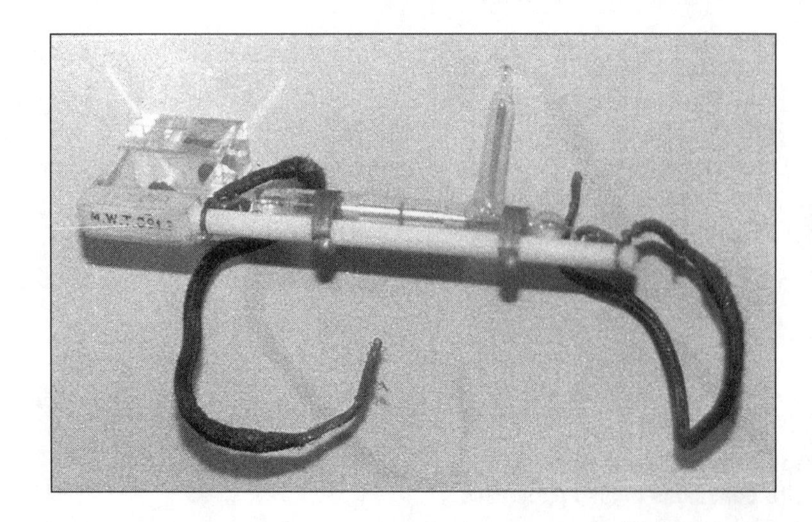

Marconi's 1898 wireless detector, called a *coherer,* was a tube filled with fine metal filings. Its electrical resistance drops when a radio signal appears across its terminals.

A replica of Marconi's 1901 transatlantic transmitter. Closing the key at the left energized the large inductive coil in the center. The coil discharged its energy as a spark across two enclosed spheres at the right. The sparking sent an electromagnetic signal through the air.

phy, he replied, "No, it would need a [reflector] the size of a continent."

Hertz was an excellent teacher with impressive language skills. He spoke fluent German, English, French, and Italian, and had a working knowledge of Arabic. His work resulted in the first long-distance wireless communication. Italy's Guglielmo Marconi (1874-1937) sent a wireless Morse code signal across the English Channel in 1899, a distance of about 30 miles. But Hertz did not live to see it happen. After contracting an infection from a decayed tooth, he died just before his 37th birthday. In 1960, in his honor, the international General Conference on Weights and Measures adopted the unit of *hertz* as one cycle/second.

Radio

Other innovators developed the technology Hertz discovered. Marconi, the first to achieve dramatic success, shared the 1909 Nobel Prize in physics with Germany's Karl Braun (1850-1918). Braun also worked on wireless communication, including early television. Although often associated with radio, Marconi actually gave the world wireless telegraphy. The word *radio* was suggested as a trademark at a 1906 international convention in Berlin. It came into wide use after 1915.

Guglielmo Marconi. Born into a wealthy family in Bologna, Guglielmo Marconi enjoyed all the things that money could buy. He received his primary education from private tutors. While growing up, he and his brother traveled extensively with their mother, staying for long periods of time in various European cities. Marconi's mother gave him the social skills that allowed him to move easily among different cultures.

At the age of 13, Marconi enrolled at the Technical Institute in Livorno. Hearing of Hertz's recent experiment with sending electricity through the air made Marconi interested in the possibility of wireless telegraphy. But his educational background proved weak, and in 1894 he failed entrance examinations for the University of Bologna. Still, he received unofficial permission to use the university's laboratory facilities.

One day in 1895, Marconi and his older brother Alfonso took a variety of experimental devices outdoors. A simple high-voltage oscillator produced sparks that Marconi hoped to receive at a distant point. His receiver was called a *coherer* from the Latin word *cohaerere,* which means "together." The coherer would conduct electricity only when it received an electromagnetic signal, and the conductivity could be read by a galvanometer's needle. Tapping the glass tube filled with metal filings after a pulse would randomize the filings and prepared the coherer for the next signal pulse.

Marconi stayed with the transmitter. Alfonso went 1-1/2 miles away with the receiver and a rifle, instructed to fire the rifle if the receiver picked up an electromagnetic signal. Marconi closed the switch, and Alfonso received the signal and fired his rifle. That experiment is often cited as a first in wireless transmission.

The Italian Ministry of Posts and Telegraph showed no interest in funding more advanced experiments. It considered wireless telegraphy useful only for ship-to-shore, or ship-to-ship, communication. They suggested that Marconi approach Great Britain, the world's most important seafaring nation at the time.

Using money from both the British government and his own family, Marconi established a research organization and sur-

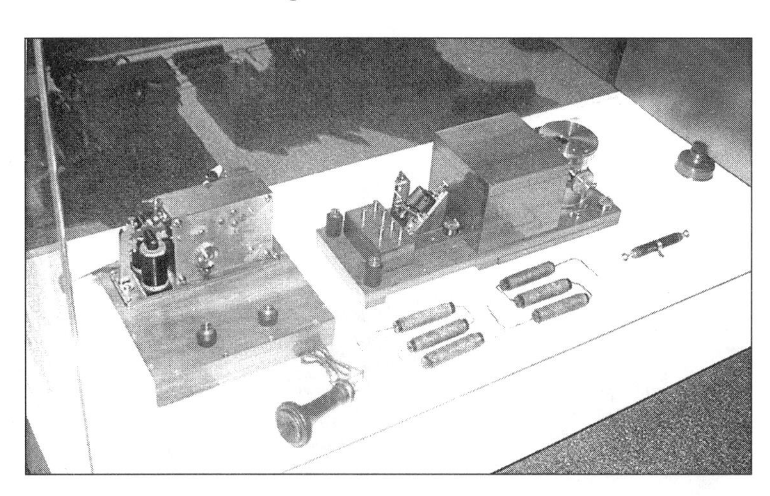

A replica of Marconi's 1901 transatlantic receiver. An antenna attached to a kite picked up the signal and transferred it to the earpiece at the left.

rounded himself with world-class technical experts. One, John Ambrose Fleming, would later invent the first vacuum tube. Marconi wanted to send and receive signals at greater and greater distances. He found that interrupting high-voltage spark signals produced Morse code that his coherer could detect. Marconi transmitted nine miles across the Bristol Channel in 1897. Then, in 1899, he went 28 miles across the English Channel. But his greatest challenge was sending and receiving electromagnetic signals across the Atlantic Ocean.

Assisted by Fleming, Marconi set up a transmitter in southwest England and a receiver in St. John's, Newfoundland, about 2,100 miles away. Marconi's lifelong friend Luigi Solari operated the transmitter. Solari sent three dots at specific times, Morse code for the letter "S." In Newfoundland, Marconi listened through a telephone earpiece. On December 12, 1901, he barely heard three clicks. He passed the earpiece to an assistant to verify the reception. The news broke and Marconi's name was suddenly known throughout the world.

Marconi married twice, first to the Irish Beatrice O'Brien and then to the Italian Maria Cristina Bezzi-Scali. He had four children. A careful dresser, almost all his photographs show him very well attired. He died of a heart attack at 63.

As important as Marconi's contributions were, he did not invent radio in the modern sense. He invented point-to-point wireless telegraphy, not voice transmission. Transmitting the modulated tones of voice and music required sophisticated amplification devices. John Fleming's (1849-1945) tube-type diode was the first step. Fleming created the field of electronics with his 1904 invention of the vacuum tube.

Battery-operated, three-valve radios using headphones were particularly popular in the 1920s. This one was manufactured in 1925.

John Fleming. In an early project he took on at London's Edison Electric Light Company, John Fleming worked on improving the carbon filaments in electric lamps. The filaments did not last long and they tended to darken the lamp's glass globe over time. Thomas Edison (1847-1931) tried to improve bulb performance by placing an extra electrode alongside the glowing filament. He observed in 1883 that when the extra electrode was connected to a positive voltage, a small current flowed between it and the filament. This came to be known as the Edison effect. The phenomenon was key to the functioning of vacuum tubes, but Edison did not see its value.

Fleming began serious research into the Edison effect in 1889. Hoping that it would contribute to wireless communication, he carried out countless experiments on transmission and reception. He had several special lamps made for his experiments. Fleming had a talent for linking complex electrical mathematics with their practical effect. He made many popular presentations to professional groups on the principles of wireless communication.

A friend of Marconi, Fleming became a technical advisor for Marconi's company in 1899. He helped Marconi design the transmitter used to send the first transatlantic radio signal to Newfoundland in 1901. Because the signal was hard to detect, Fleming set about improving radio circuitry. He first faced the problem of converting a weak alternating current into a direct current to operate a receiver. Fleming realized that his 1889 Edison effect lamp could convert ac to dc by allowing current to flow in only one direction. He experimentally confirmed in 1904 that his tube-type diode would detect radio waves. The electron had just been discovered and Fleming knew that the Edison effect was caused by the emission of electrons from a heated filament. Fleming's invention and its circuitry represent the beginning of the electronic era.

Fleming's diode had two electrodes inside a glass globe, which allowed current to flow in only one direction. Fleming called it a

This device was displayed in a museum under the general heading of "Some Fessenden Items." It appears to be an early radio tuner from about 1910.

valve, since it turned on when current flowed in one direction and turned off when it flowed in the other direction. Britons still call the devices "valves," but they are known as *vacuum tubes* in America. The Fleming valve allowed more precise detection of radio waves and it saw heavy use in the early years of amplitude modulated (AM) radio communication. It was an essential part of all transmitters and receivers for over 50 years.

Fleming strongly supported John Logie Baird in his effort to establish the new technology of television. He served as president of the Television Society from 1930 until his death at 95.

Fleming's valve ushered in the field of practical electronic communication. The first person to use Fleming's technology to send wireless voice and music was Reginald Fessenden (1866-1932). Fessenden sent a modulated signal, over a distance of several hundred miles, from Massachusetts to ships at sea in 1906.

Reginald Fessenden. Born in East Bolton, Quebec, in Canada, Reginald Fessenden was the oldest son of a minister. His mother, Clementina, was the driving force behind the establishment of Empire Day, a legal holiday in Canada honoring Queen Victoria that is now called Victoria Day. Fessenden's parents expected him to follow a career in the ministry or a related field.

Fessenden's interest in technology was influenced at the age of 10 when he saw Alexander Graham Bell demonstrate a telephone at his lab in Brantford, Ontario. Fessenden found an outlet for that interest through mathematics. He attended Bishop's College and, partly to please his parents, followed a program that emphasized the classics. But he had little interest in that field and left at 18 to serve as headmaster of a small private school in Bermuda. Fessenden never graduated from college.

At the school in Bermuda, Fessenden discovered *Scientific American* magazine, which he read avidly. He took a particular interest in electricity and the successes of his early hero, Thomas Edison. While in Bermuda, Fessenden met Helen Trott, whom he married in 1890. They had one son.

Fessenden eventually decided to pursue a technical career. In 1885, he took a job at Edison's Pearl Street electric plant in Manhattan. Later, he transferred to Edison's research lab in Orange, New Jersey. He next worked for a subsidiary of Westinghouse Electric and Manufacturing Company in Newark. Then, he used his work experience to move into electrical engineering professorships, first at Purdue University and then at the University of Pittsburgh. At both universities, he conducted introductory investigations into wireless communication.

In 1900, the U.S. Weather Bureau asked Fessenden to develop a wireless system for distributing forecasts and relaying meteorological data. The Weather Bureau offered Fessenden $3,000 per year, assistants, workspace, and housing at Cobb Island, Mary-

A replica of a Reginald Fessenden electrolytic liquid detector. In 1906, Fessenden used one like this in the first public broadcast of the human voice.

Another "Fessenden Item." Its purpose is hard to determine because its unusual appearance is not typical of other early radio equipment.

land. Drawn by the opportunity to work full time as an inventor, Fessenden accepted the challenge.

Fessenden and his assistants developed a workable spark-gap system that transmitted Morse code over a distance of 50 miles. During this period, he grew particularly interested in sending sound by wireless methods. The idea of transmitting a continuous carrier frequency, while varying the amplitude of the wave, occurred to him. The variations could follow the irregularities of sound, particularly speech, and sorting the variations at a receiving station could reconvert them into intelligible speech.

Fessenden tested higher frequencies and tried to improve the sensitivity of his components. He knew that these changes could help him transmit continuous waves instead of the unmodulated, intermittent dot-dash tapping of Morse code. In December 1900, Fessenden sent a poor-quality voice signal between two 50-foot tall wooden masts positioned one mile apart. This was the first radio transmission of voice.

Fessenden used a coherer like Marconi's, a glass tube filled with metal filings that were affected by electromagnetic radio signals. To operate properly, the coherer required continual tapping with a vibrator to reorient the filings for each new part of the wireless signal. It distorted speech and was not sensitive enough for routine sound transmission. Fessenden developed a detector that would more effectively read a continuous signal. He called his design, which basically consisted of two platinum wires with their ends dipped in a small pool of nitric acid, a *barretter*.

The system worked well enough that Fessenden secured financing to establish the National Electric Signalling Company (NESCO) in about 1902. Serving as company president, Fessenden aimed to produce a workable wireless voice communication system. As part of that effort, NESCO constructed experimental stations at Machrihanish, Scotland, and at Brant Rock, Massachusetts, not far from Plymouth. Fessenden moved his family to the Massachusetts site in 1905 and immediately began setting up circuitry. Soon, he could send and receive voice messages from Scotland.

Fessenden's most important piece of equipment was a high-frequency alternator from General Electric that operated at 50,000 Hz. Typical spark-gap equipment operated at 60 Hz, which was too low to produce audible, undistorted speech. Using a 420-foot tall antenna, on Christmas Eve 1906, he sent the world's first public wireless signal of voice and music. Fessenden hosted a program at 9 P.M., playing musical records and talking. An accomplished amateur musician, he also played "O Holy Night" on a violin and sang the verses. He asked listeners to send him letters reporting on the broadcast. The response confirmed Fessenden's success at inventing radio transmission. Several ships owned by the United Fruit Company reported receiving the broadcast from their positions on the Atlantic Ocean. Others heard it as far away as the West Indies.

That 1906 broadcast was the high point of Fessenden's career. NESCO went bankrupt in 1911 and Fessenden left radio behind him. He found himself a failure at both business and self-promotion. Yet both were of critical importance to financial success in early wireless broadcasting. Many new and powerful radio organizations used several of Fessenden's innovations without permission, and trying to defend his patents exhausted him. He chose to move on to work with turboelectric drives for ships, underwater sending and receiving equipment, depth finders, and other devices. He found moderate success in those endeavors and accumulated a lifetime total of over 500 patents.

Late in life, Fessenden won a $2.5 million out-of-court settlement from the Radio Corporation of America for patent infringement. He and his wife used the money to purchase property on Bermuda, where he lived for the last four years of his life. Fessenden died in Bermuda at the age of 65.

Fessenden's accomplishments proved pivotal to the early success of voice transmission. Others added improvements to make radio available to more people. One addition was the vacuum tube amplifier invented by Lee De Forest (1873-1961). Its importance to early wireless voice communication cannot be overemphasized. Historian Mitchell Wilson described De Forest's 1907 audion as an "invention on a magnitude that may be touched only two or three times in a century."

Lee De Forest. After receiving a doctorate from Yale University, Lee De Forest spent three years at Chicago's Illinois Institute of Technology, where he worked on experimental communication devices. He then persuaded a group of investors to establish the American De Forest Wireless Telegraphy Company. The company received further support from the U.S. government and eventually built 90 stations.

Wanting to create a more sensitive detector, De Forest placed a zig-zag-shaped length of nickel wire between the anode and cathode of a diode. The resulting three-element tube detected Morse code far better than any diode and produced dot-dash sounds. De Forest called his tube the *audion*, or *triode*, and took out a patent in 1907. It served as the prototype for billions of radio amplifier tubes that followed.

De Forest did not at first understand the triode's capabilities. Five years later, he realized its most important feature: It could not only act as a detector but could also amplify voice. The output of one triode could be connected to the input of another, and by using this staging technique, several triodes could amplify weak signals. This allowed for the use of speakers, rather than cumbersome headphones, for radios. The introduction of speakers made radio marketable to the general public.

The stakes were enormous. In November 1920, KDKA in Pittsburgh became America's first commercially licensed radio station. Transmitting with a power of 100 watts, the station's first programming aired the Warren Harding/James Cox presidential election returns. Just 10 years later, 14 million radio sets were in use. That number grew to 44 million in 1940. Other inventors soon claimed they had previously discovered De Forest's unique circuitry. One legal claim—between De Forest and Edwin Armstrong—resulted in one of the longest patent battles in history. After 14 years of legal battles, the Supreme Court decided in 1934 in De Forest's favor. During those protracted proceedings, De Forest's audion made fortunes for others, helped create RCA, and brought a new form of entertainment to the public. But De Forest received little wealth or fame, and much unhappiness, from his invention.

Whenever De Forest tried to manufacture or market his inventions, he ran into lengthy patent challenges. In his 50s, De Forest wisely changed his strategy. He simply sold his patent rights to others.

Radio communication has many unsung heroes. Another was Edwin Armstrong (1890-1954), who single-handedly developed the sophisticated circuits that form the basis of almost all modern radio and television receivers. But Armstrong's greatest accomplishment was inventing static-free FM radio transmission and reception equipment. His frequency modulation principle is used not only in radios and televisions, but also in telephone, radar, and spacecraft communication networks. Armstrong's very active life was filled with discovery, financial suc-

Speakers were not practical for early battery-powered radios. Large horns often substituted, as in this 1925 RCA Model 3 Radiola.

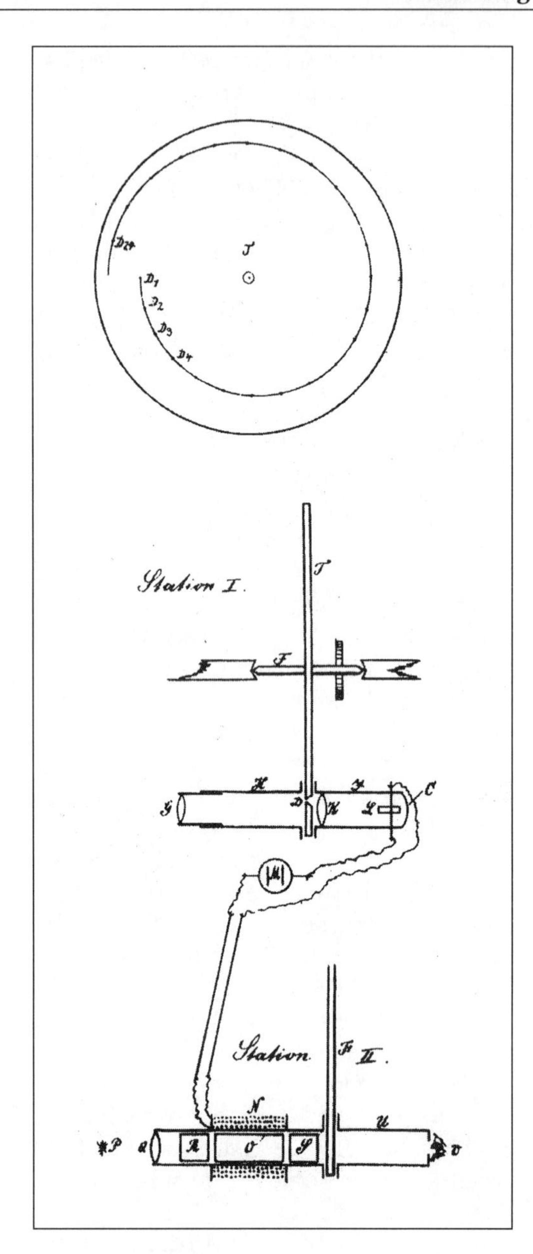

Drawing from an 1884 German patent by Paul Nipkow. The top figure shows the spiral arrangement of the small holes used in mechanical television scanning. Loosely translated, the patent title is "Electric Potential for Viewing at a Distance."

cess, court battles, and ultimately, tragedy.

Edwin Howard Armstrong. As a teenager growing up in Yonkers, New York, Armstrong read Marconi's *The Boy's Book of Invention* and set about becoming an inventor in the emerging field of wireless communication. He filled his room overlooking the Hudson River with crystals, coils, condensers, and resistors. After high school, he commuted by motorcycle to Columbia University, where he earned a degree in electrical engineering. One of his teachers was Michael Pupin (1858-1935), developer of a radio tuning circuit and other communication inventions.

Under Pupin's influence, Armstrong made his first significant discovery while he was still in college. He devised a regenerative circuit for use with De Forest's new audion tube. During his senior year at Columbia, Armstrong built the circuit and found that it greatly improved radio reception. Conducting research at Columbia after graduation, he worked on a series of circuits that increased the sensitivity of radio receivers. His work brought him in contact with RCA, which paid him quite well for his patent rights. While in his 30s, he became a millionaire. In 1923, he married Marion MacInnis, the secretary of RCA's president.

As the 1920s wore on, Armstrong found himself trapped in a corporate war over control of radio patents. RCA, Westinghouse, American Telephone & Telegraph, and others all wanted to build large corporations based on different people's patents. Patent infringement cases were common during that time. When he lost the Supreme Court decision to De Forest in 1934, it was a heavy blow to Armstrong and his supporters. Many people thought that the Supreme Court based its judgement on a judicial misunderstanding of the technical facts. In a show of support, the Franklin Institute weighed all the technical evidence and awarded Armstrong the highest honor in U.S. science, the Franklin Medal.

Signal interference associated with AM transmissions posed a serious problem during radio's early years. Unlike most others, Armstrong felt that frequency modulation (FM) was the only long-term solution for eliminating static. RCA, which had heavy investments in AM transmitters and receivers, asked Armstrong to remove his FM equipment from the space RCA provided for him in the Empire State Building. He moved to a large apartment overlooking the East River and personally financed his work. Armstrong field-tested his experimental equipment in 1933 during a violent thunderstorm and received static-free sound from 80 miles away. In July 1939, Armstrong began broadcasting from the world's first FM station, W2XMN, which he built in Alpine, New Jersey.

RCA brought out its first FM receiver shortly after World War II ended. It was an obvious adaptation of Armstrong's patented circuit, and Armstrong filed a suit against RCA. When RCA dragged out the proceedings, Armstrong's legal fees mounted steadily. The financial and emotional strain was more than he could bear. He took his own life in 1954. Shortly after that, his widow won $10 million in damages in 21 patent infringement suits. The proceedings involving Armstrong's innovations didn't close completely until 1967, when the Supreme Court refused to review a lower court judgement against Motorola Corporation.

The 1929 Baird Televisor had a rotating disk at the back. Its speed was synchronized with the disk used in the camera at the television studio. The television's small viewing screen is at the right. In front of the set is an early cathode-ray tube used for all-electronic television.

Television

Technologists began investigating television transmission even before electronics was discovered. The idea was to break an image into small sections and send pulses down a wire. France's Constantin Perskyi first used the word *television* in August 1900 at an international electricity conference in Paris.

The first moderately successful transmission and reception was based on a spinning disk. The disk had several small holes in a spiral near its edge and rotated at about 600 rpm. The method was called *mechanical television,* and John Logie Baird (1888-1946) developed the first practical system. The British Broadcasting Corporation granted him the world's first television license in 1929.

John Logie Baird. Born in Helensburgh, Scotland, just across the River Clyde from James Watt's birth city of Greenock, John Logie Baird was the son of a poorly paid minister. The youngest of four children, Baird endured respiratory ailments all his life. After secondary schooling, he went on to study electricity at the Royal Technical College in Glasgow. He later held several different jobs but often missed work because of chronic illness. Bad luck in many areas dogged Baird through his early years. At 34, sick, with no job and little money, he felt like a failure.

Baird then moved in with a childhood friend who lived in Hastings, England, and his health slowly improved. Recalling discussions of television at school and his youthful interest in electricity, he decided to try and make a practical television system. Success at that goal had eluded all other inventors.

In the early 1920s, investigators used various forms of the Nipkow disk as an experimental transmitter. Named for Germany's Paul Nipkow (1860-1940), it was a revolving disk that had holes in a shallow spiral near the edge. The holes scanned a subject, separating the image into smaller units. Tapping into his small savings, Baird assembled experimental equipment using Nipkow disks in both transmitter and receiver. In 1924, he transmitted a crude image of a Maltese cross over a distance of a few feet. This limited success encouraged him to take the apparatus to London.

Baird found a financial backer who of-

This incomplete 1928 No. 1 Jenkins Home Radio Vision Kit has a rotating disk with small holes. The holes mechanically scanned the image and transferred the information to a photocell just visible at the top right.

Library of Congress

C. Francis Jenkins is shown here in 1928 holding film that he planned to broadcast with his experimental mechanical television equipment.

fered a small amount of money and an apartment that he could use for a laboratory. More success came in 1926 when Baird demonstrated the transmission and reception of people's faces to a technical audience. The faces appeared as flickering pinkish images on a four-inch by two-inch screen. Baird became famous overnight. Subsequent publicity brought investment money and he was never poor again.

The BBC granted Baird a six-year experimental license in 1929 to transmit television signals using his mechanical scanning system. He used 2TV for his station's call letters. The public purchased about 20,000 of his receiving sets, called Baird Televisors. Programming in the modern sense did not exist. The fuzzy images people saw in their homes were often just different types of test transmissions. Baird operated color equipment in 1928 and sent a transatlantic television signal the same year. Still, Baird was unwilling to acknowledge the considerable benefits of electronic television and some people experienced him as quite abrasive.

Baird married concert pianist Margaret Albu in 1931 and they had two children. He died at 58 in Bexhill, England.

Baird was not the only person working in the field of mechanical television. Another, Charles Francis Jenkins (1867-1934), was America's most successful early television pioneer. The Federal Radio Commission granted him a radio license in 1928, which he used for experimental television transmission. Operating from Washington, D.C., its call letters were W3XK.

Charles Francis Jenkins. Born near Dayton, Ohio, C. Francis Jenkins spent his boyhood on a farm near Richmond, Indiana. After graduating from nearby Earlham College, he moved to Washington, D.C., in 1890. Jenkins went to work for the Life Saving Service, now the U.S. Coast Guard. Impressed by Thomas Edison's work with motion pic-

Vladimir Zworykin's 1920s flying beam iconoscope is the basis for all image pickup tubes used in television cameras. This one is displayed in front of a photo of Zworykin holding an iconoscope.

ture equipment, he spent his free time constructing an improved film projector. He completed an experimental one in 1894, which he called the Phantascope. He was barely 28 when he left his job with the Life Saving Service for a career as an inventor.

Jenkins teamed up with Thomas Armat, and together they built the first practical motion picture projector. The two men received a joint patent in 1897 on a design that has influenced all modern projectors. Edison bought the patent rights, added his own improvements, and marketed their projector as the Vitascope. Soon after, Jenkins and Armat went their separate ways. Jenkins married Grace Love in 1902.

To help establish film standards, Jenkins organized the Socicty of Motion Picture Engineers in 1916. His motion picture work led him to consider television. Jenkins received support from the U.S. Navy and used navy facilities at Anacostia, about five miles from Washington, D.C. One of his mechanical television designs used a rotating Nipkow disk with 24 small holes. The disk was sometimes used with a large lens that focused reflected light from the subject onto photocells behind the scanning disk. The cells converted the image into a pulsing electric current. Jenkins's first synchronized wireless transmission of moving images came in 1925. He transmitted a motion picture of a rotating windmill from Anacostia to Washington, D.C. The Secretary of the Navy, the Secretary of Commerce, and other government officials viewed the 48-line, low-definition image. Jenkins used two different terms to describe the emerging technology: *radio vision* and *home radio movies.* The word *television* had not yet come into common use.

After receiving his wireless transmission license in 1928, Jenkins began experimental, once-a-day, half-hour television broadcasts in 1930. Programming expanded to four hours a day the next year. But the crude flickering images tended to look like silhouettes. The De Forest Radio Company purchased the Jenkins Television Corporation in 1929. Jenkins kept his research laboratory but soon resigned because of ill health. He was a prolific inventor who eventually held 400 U.S. and foreign patents. He received medals of recognition from the Franklin Institute and the city of Philadelphia. Jenkins wrote five books on early television before his death in 1934.

Mechanical vs. Electronic Television. One major problem with the mechanical rotating-disk system was the poor quality of the images it transmitted. This came partly from the disk's inability to spin quickly enough. A good mechanical system might achieve 60 lines of resolution at 18 frames/second—not good enough to look natural. Electronic methods were trickier, but they produced much better results. By the mid-1930s, they produced hundreds of lines of resolution at 30 frames/second.

A close-up view of Zworykin's iconoscope

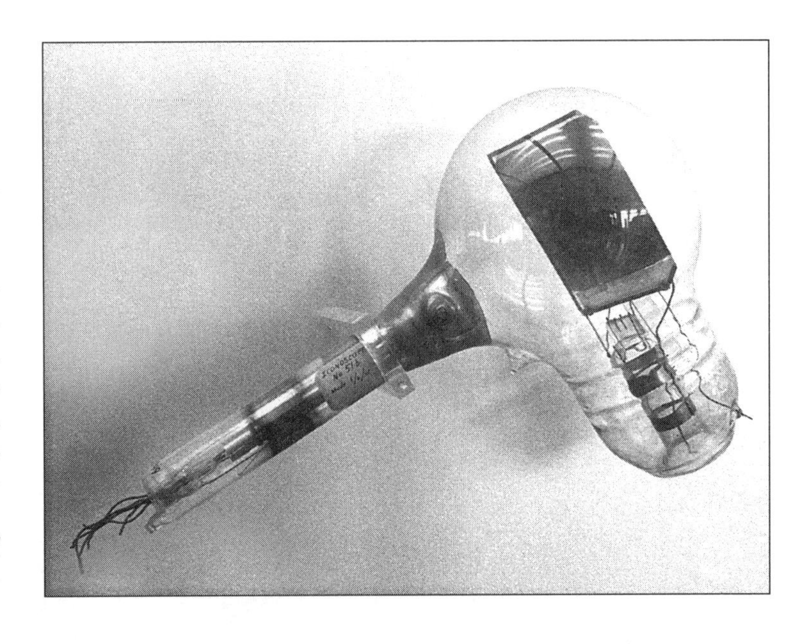

This museum display shows television image resolutions. The top one has 60 lines as in 1931 televisions; the middle has 180 lines as in 1934 sets; and the bottom has 625 lines as in sets made in 1948 and later. (Visible in the background are the twin steeples of Munich's St. Michael's Cathedral.)

The Nipkow disk showed technologists one way to send television images. However, being a mechanical system, it could not scan quickly enough to transmit and receive high-quality moving images. Electronic systems did not suffer that shortcoming. Properly set up, an electronic scanner and receiver could operate at the speed of light.

In spite of their speed, electronic television systems still relied on the logic provided by mechanical Nipkow disk. Cameras scanned objects with a rapidly moving electronic beam. The process worked the same way that a person's eye reads a printed page. The beam moved from left to right across the image recording dots and empty space, then returned rapidly to the left and scanned the next line. Early experimental systems scanned 343 lines. Transmitting equipment sent the pulses of dots along a wire or through the air. Electronic receivers rearranged the pulses and displayed them on a television screen.

At least that was the idea. It took a long time to travel from theory to practice. The invention of De Forest's and Fleming's vacuum tubes encouraged early investigators to work on electronic television. Developing a flying beam for cameras proved a complex task, so early electronic television pioneers concentrated their efforts on amplifying the output. Boris von Rosing (1869-1933), a professor at the St. Petersburg Institute of Technology in Leningrad, Russia, took that route. Von Rosing and his young assistant Vladimir Zworykin (1889-1982) experimented with primitive electronic amplifier tubes.

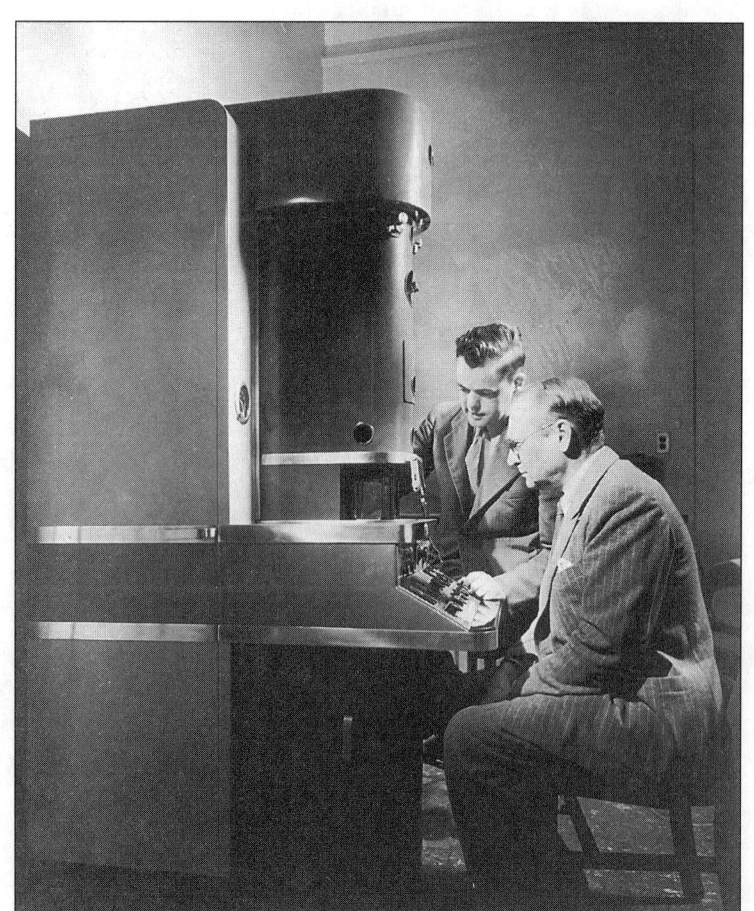

Library of Congress

Vladimir Zworykin (right) and James Hillier worked together to develop a practical electron microscope in the 1940s. This photo was taken sometime before 1950.

Vladimir Zworykin. Born in Murom, Russia, about 200 miles east of Moscow and near the Oka River, Vladimir Zworykin was the youngest of seven children. His father owned and operated a fleet of riverboats

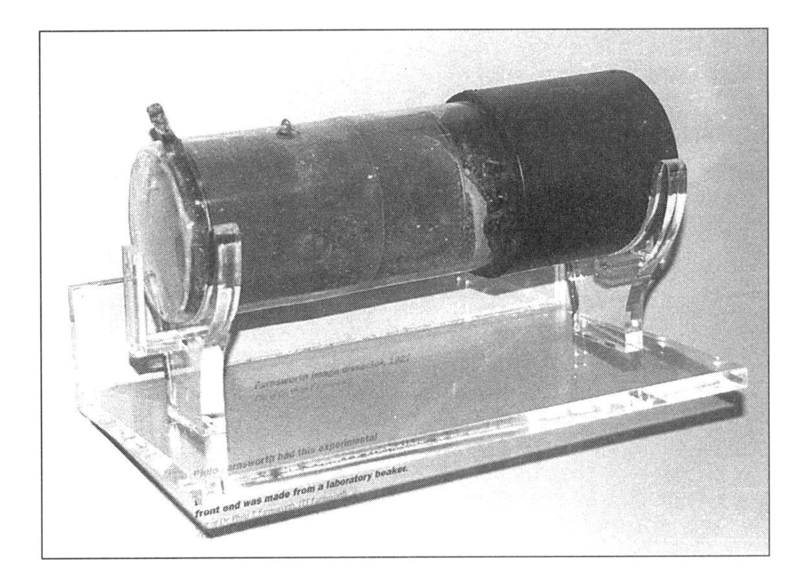

Philo Farnsworth made this experimental 1927 dissector from a laboratory beaker.

and the young Zworykin helped him during his school vacations. He learned basic electricity on the journeys by reading books and repairing electrical equipment. After graduating from high school in 1906, Zworykin decided to study electrical engineering at the St. Petersburg Institute of Technology.

Zworykin soon met von Rosing, who was working on a method to send a picture by wire. He was the first person to attempt transmission of an image by scanning the inside of a cathode-ray tube. Von Rosing freely allowed Zworykin to assist him in his research. He felt the future of television was in the cathode-ray tube, not in the mechanical systems being investigated by others. Zworykin stayed on after graduation to assist von Rosing for several months.

Then, during the chaos of the 1917-1921 Russian Revolution, Zworykin wandered for months to avoid arrest by competing armies. Making his way to the port city of Archangel, he obtained a visa that allowed him to sail to America. He reached New York City in 1919. He found a job with the Westing-house Electric and Manufacturing Com-pany's research labs in Pittsburgh. His first assignment involved work on new radio tubes and photoelectric cells. Zworykin became an American citizen in 1924, the year he first demonstrated a television system.

Zworykin had applied for a patent the previous year for his *iconoscope*, or television transmitting tube. The ancient Greek word *eikon* means "image" and *scopus* means "watcher." Zworykin showed Westinghouse executives the first flickering images from his experimental system. That same year, he patented his *kinescope*, or television receiving tube. The ancient Greek *kinema* means "motion." The two components set the stage for a practical television transmitting and receiving system.

Unable to convince Westinghouse executives of the worth of his invention, Zworykin moved to RCA in 1929. He served as director of RCA's electronic research laboratory in Camden, New Jersey. Later that year, Zworykin demonstrated the first all-electronic experimental television system at a convention of the Institute of Radio Engineers. RCA went on to spend millions of dollars to develop a practical system for the public. In 1936, it provided television sets for 150 New York City area homes and began experimental telecasts. The National Broadcasting System, a division of RCA, established regular telecasts in 1939. Zworykin's inventions and specialized circuits found such universal use that he is often described as the inventor of television, though he always rejected the title.

After the iconoscope-type television system, Zworykin's most important work was with electron microscopes. He and James Hillier (1915-) teamed together during the early 1940s to create electron microscopes that magnified as high as 200,000 times, when optical microscopes had a limit of 2,500 times. Zworykin retired from RCA in 1954, though he kept an office at RCA's laboratories. In 1966, President Lyndon Johnson awarded him the National Medal of Science, America's highest scientific honor. Zworykin and his wife, Katherine, kept an open house in Princeton for their 5 children and 17 grandchildren. Zworykin died one day before his 93rd birthday.

An image of Philo Farnsworth appeared on a 1983 commemorative postage stamp, along with the words "First television camera."

Like this one from 1936, some early all-electronic televisions had mirror lids. Picture tubes of that era were so long that they were sometimes mounted vertically.

Zworykin's iconoscope, with its electronic flying beam, was a remarkable invention. Practically all large technical museums now have one on display. But it was not the only way to get the job done. Philo Farnsworth (1906-1971) used a related device that he called a *dissector*.

Philo Farnsworth. Philo Farnsworth, called Phil by nearly everyone, was born into a farming family that lived near Beaver, Utah. When he was young, his family moved into a house in Idaho that had back issues of technical publications that the previous occupants had left behind. Farnsworth read them all and grew fascinated with the possibility of using photoelectric cells and cathode-ray tubes to transmit images through the air. At the age of 15, he began designing his own system.

Later, when Farnsworth worked for the Community Chest in Salt Lake City, two co-workers took an interest in his television ideas. With a $6,000 investment from them, Farnsworth moved to Los Angeles where he rented space for his experimental work and an apartment with his new wife, Elma. The couple eventually had four children.

Living in Los Angeles brought Farnsworth into contact with others working in the emerging field of television. But he was barely 20 years old, had attended college for only two years, and often found it hard to convince others that he was a serious researcher.

The key innovation in Farnsworth's design was the camera tube. He called it a *dissector*, because it dissected an image into bits. Its glass tube had a photosensitive mirror at one end on which the image was focused. The mirror emitted electrons in response to the different light intensities projected on it. Farnsworth first demonstrated the entire system to his financial backers in 1928, one year after he applied for a patent. The crude image was good enough to help him establish a small production company named Television, Incorporated. RCA and other companies had invested large sums of money in the Zworykin system and did not welcome Farnsworth's competition.

Always just barely keeping his company solvent, Farnsworth made his first open public demonstration of electronic television in 1934 at Philadelphia's Franklin Institute. To throw Farnsworth off guard, RCA used

Because of their large size, early television sets were treated as separate pieces of furniture. This unlabeled console television was probably made in the early 1950s and had a screen size of about 6 inches.

The Dumont Corporation was an early manufacturer of electronic televisions. This 1939 Dumont cost $395. One of the company's television picture tubes is displayed on top of the set.

its influence to persuade the Federal Communications Commission to avoid establishing television on a nationwide scale. But Farnsworth had 165 patents and legally controlled several crucial aspects of broadcasting. By 1939, RCA had spent $9 million developing a version of electronic television. Farnsworth had spent about $1 million at his Fort Wayne, Indiana, plant and the end was not in sight. He agreed to sell his patent rights to RCA that year, an action that effectively removed him from active research in television.

Farnsworth was described as an intense, but likable, person. He met with Zworykin several times. After selling his patent rights to RCA, he enjoyed spending time at his 90-acre farm in western Maine where he had a small laboratory. He died at 64 in Salt Lake City.

Early in the history of television, no one foresaw the control of electronic communication by a few huge corporations. Independent inventors helped launch television, but many were forgotten in the dynamics of the national enterprise. Had he not worked outside the technical mainstream, Farnsworth might be better remembered as a television pioneer. Still, only one pioneer of radio or television equipment made *Time* magazine's list of the 100 Most Important People of the 20th Century: Philo Farnsworth.

Magnetic Tape Recording

To succeed, commercial television required a host of new technologies, including the development of a high-quality recording device. Recording was particularly important in America because many television shows originated in New York City, three time zones away from the heavily populated West Coast. An early solution involved filming a picture from a high-resolution television set as a program was performed live in New York City. After quick processing, the film was later rebroadcast for viewers in the west. The method proved expensive and the picture quality poor.

Several manufacturers began to look into the possibility of using magnetic tape for video recordings. Audio recordings on tape had grown popular in the late 1940s. Three organizations pioneered the technology: RCA, Bing Crosby Enterprises (BCE), and Ampex. Rising to the top of the pack, Ampex went on to produce the first practical videotape player in April 1956.

Audio Recording. America's entry into audio recording started with Major Jack Mullin (1915-1999), a member of the Army Signal Corps during World War II. Sent to Frankfurt as the war ended, Mullin saw two of Germany's secret audiotape recorders at a radio studio. The recorders, called Magnetophons, used a recording tape coated with iron oxide. Mullin took two Magnetophons back to America along with 50 reels of audiotape.

Mullin made his first demonstration in America to stunned members of the Institute of Radio Engineers at their 1946 convention in San Francisco. Singing star Bing Crosby immediately got involved, signing Mullin up as the chief engineer of BCE. The season premier of Crosby's radio show in October 1947 served as America's broadcast premier of magnetic tape recording. The reproduced music sounded so much like a live broadcast that other radio stars immediately began to prerecord their shows. And audiotape technology provided the basis for videotape recording.

Early investigators of videotaping soon discovered that video signals required

The 1964 Philips Model EL-3400, made in the Netherlands, was one of the first videotape recorders to use a helical scan system. It used one-inch reel-to-reel tape and had a maximum running time of 45 minutes.

much more space than audio signals. An audiotape running at the standard speed of 15 inches/second would have to operate at 400 inches/second for video. And countless other technical problems arose. Many people at several companies worked to solve them. The first to succeed worked for Russian-American Alexander Mathew Poniatoff (1892-1980). Poniatoff used his own initials, plus "ex" for "excellence," in the title of a company he formed in 1944, Ampex.

Alexander Poniatoff. Born in the Kazan District of Russia, about 450 miles east of Moscow, Poniatoff was the son of a prosperous lumber industry executive who made sure he had a good education. The younger Poniatoff earned degrees in electrical and mechanical engineering from German and Russian universities. During World War I, Poniatoff served as a pilot in the Imperial Russian Navy. Like Vladimir Zworykin, he found himself caught up in the 1917-1921 Russian Revolution, where his skills as a pilot found particular demand. After escaping to China in 1920, Poniatoff immigrated to the United States in 1927. He became an American citizen in 1932.

Poniatoff first found work in the research and development labs of the General Electric Company in Schenectady, New York. Later, he moved to the Dalmo Victor Company in Belmont, California. During World War II, he was involved with airborne radar for the U.S. Navy. Poniatoff formed Ampex in Redwood City, California in 1944. The company's main purpose was constructing components for radar systems developed by GE. Its products included precision-tuned motors, which later became important in Poniatoff's audiotape drive systems.

After the war, Poniatoff switched his company's focus to producing audiotape players like those developed in Germany under wartime secrecy. He received support from Hollywood's BCE, which needed high-quality replay facilities for radio-broadcasting purposes. With Mullin as an occasional consultant, Ampex produced the first American-built magnetic audiotape player in 1948. The Model 200 used an improved tape developed by the 3M Corporation.

Ampex also produced the first data instrumentation recorder for storing large amounts of information on magnetic tape. Ampex machines were used in laboratories and onboard airplanes to record rapidly generated scientific information. Specializing in the manufacture of professional-quality equipment, the company prospered in the late 1940s. But like others, Poniatoff saw a large market presenting itself in the field of video recording. The potential was particularly tempting in America because of the time shifting required for broadcasting over a large country's time zones.

By the late 1940s, teams of researchers who pooled their knowledge were completing important technical work. Established in 1951, Ampex's videotape recorder team ultimately consisted of six members headed

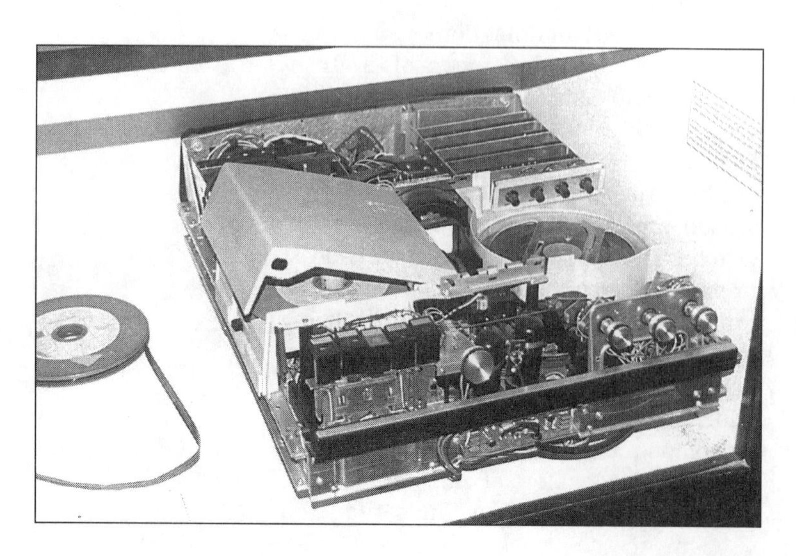

CBS offered this video Teleplayer for sale in 1971. Customers purchased feature films on 8 mm tape to view through their televisions.

by Charles Ginsburg (1920-). Ginsburg had diabetes and was one of the world's earliest users of insulin. Another team member, Ray Dolby (1933-), would go on to develop the Dolby noise-reduction system used in audio equipment.

Minor successes followed failures as the Ampex team put in long hours of work over many months, which eventually grew into years. The team's final design was a four-head, transverse-scan system in which the heads rotated at 14,400 rpm across a two-inch-wide tape. The team's key innovation was the idea of recording on a bias, a strategy like that of slicing carrots on a slant to expose more surface area during cooking. After months of hectic development work, Ampex demonstrated its videotape player to about 200 CBS managers at Chicago's Hilton Hotel in April 1956. CBS's vice-president of engineering made a presentation with black-and-white television monitors lining the walls to make him visible to everyone present. Unknown to the audience, he was also being videotaped. As the presentation was played back to the puzzled managers, a curtain was drawn to expose the Ampex recording equipment. The audience applauded wildly for this first successful demonstration of a practical videotape player. Ampex processed $5 million worth of orders over the next four days. Its first production videotape recorder was the desk-sized VRX-1000, which had its initial broadcast use with a November 30, 1956, news show.

Poniatoff served as president of Ampex until 1955, when he became chairman of the board. By the time of his death at age 88, the company did business worth a half-billion dollars each year.

References

"Pioneers of Television—Charles Francis Jenkins" by Albert Abramson, *SMPTE Journal* [a publication of the Society of Motion Picture & Television Engineers], February 1986.

Electronic Motion Pictures by Albert Abramson, Arno Press, 1974.

Pioneers of Electrical Communication by Rollo Appleyard, Books for Libraries Press, 1968.

Asimov's Biographical Encyclopedia of Science and Technology by Isaac Asimov, Doubleday Publishers, 1964.

The Communications Miracle by John Bray, Plenum Press, 1995.

Marconi, The Man and His Wireless by Orrin E. Dunlap, Macmillan Publishers, 1937.

The Story of Television—The Life of Philo T. Farnsworth by George Everson, Arno Press, 1974.

John Logie Baird and Television by Michael Hallett, Priory Press, 1978.

Marconi by W. P. Jolly, Constable and Company Publishers, 1972.

Empire of the Air by Tom Lewis, Harper Collins Publishers, 1991.

"Radio Revolutionary" by Thomas S. W. Lewis, *American Heritage of Invention and Technology*, Fall 1985.

Those Inventive Americans, National Geographic Society, 1971.

"Agents of Change" by Phil Patton, *American Heritage*, December 1994.

American Science and Invention by Mitchell Wilson, Bonanza Books, 1960.

"The Race to Video" by Stewart Wolpin, *American Heritage of Invention and Technology*, Fall 1994.

"The Forgotten Father of Radio" by William S. Zuill, *American Heritage of Invention and Technology*, Summer 2001.

DIVIDER AND SQUARE ROOTER
THE ENIAC

CHAPTER 5
Computers

Introduction

The Book of Lists, by David Wallechinsky and others (William Morrow Publishers) was first published 1977. Its appearance triggered a widespread interest in various types of lists. When some mid-21st century author makes a list of the most significant 20th century inventions, the digital computer will surely be in the running.

Contemporary computer professionals cite Charles Babbage as the first in their field. Babbage designed the essential parts of a digital computer before Queen Victoria took the throne of his native England. One of his tasks involved developing a way to communicate with the machine. Working with Ada Lovelace, he decided to use punched cards, a method France's Joseph Marie Jacquard had devised in 1804. Jacquard's cards controlled the weaving patterns produced by his specialized textile looms.

Howard Aiken and Grace Hopper, who formed another team not unlike that of Babbage and Lovelace, worked together in America on the world's first program-controlled digital computer. The Mark I went operational in the early 1940s. Although Aiken and Hopper didn't know it, Germany's Konrad Zuse worked on a similar device around the same time.

Improvements to the computer mushroomed after World War II. ENIAC, the first electronic computer, followed the electromechanical Mark I in 1946. J. Presper Eckert and John Mauchly headed its design team. Remington Rand Corporation manufactured the first commercially successful computer, the 1951 UNIVAC.

The computers of the 1950s and 1960s were far too expensive for personal use. Stephen Jobs and Steven Wozniak addressed that problem in 1977 by producing the first widely accepted personal computer (PC), the Apple II. Another computer team, William Hewlett and David Packard, started the Hewlett-Packard company. William Gates and Paul Allen established the Microsoft Corporation in 1975. These three sets of partners helped to provide the inexpensive computers and software that contributed to the birth of the Internet in 1983.

Prehistory of Computers

One of the most basic concepts of computer technology deals with method of communication—transferring commands from the operator's mind to the computer. The approach used for over a hundred years required punched cards or tape. French weaver Joseph Jacquard (1752-1834), who wanted a reliable method for duplicating complex textile patterns, devised it.

Joseph Marie Jacquard. Born in the city of Lyon, Joseph Marie Jacquard learned silk weaving at an early age from his father. As a teenager, Jacquard apprenticed himself to a bookbinder and went on to work at various trades until he inherited the family home when his parents died. Jacquard then returned to weaving.

The Smithsonian Institution now displays a portion of the 1946 ENIAC computer.

Joseph Jacquard's 1804 automated textile loom involved the first mechanized process to use punched cards. The many thin wooden cards looped at the left controlled the pattern of the colored threads in the fabric.

Hand methods often produced poor-quality patterned silk fabrics and Jacquard wanted to automate the weaving process. Silk weaving involved raising and lowering lengthwise threads, called the *warp*. A shuttle passing between the warp carried the *weft*, or crosswise threads. Each individual weaver developed a pattern by raising different groups of warp threads and changing the colors of the weft threads. In one weaving method of Jacquard's time, warp threads passed through hooks in long, thin metal rods. An apprentice had to sit on top of the loom, lifting or lowering the rods in a precise order. An apprentice losing track of the sequence resulted in a poorly made pattern.

Jacquard came to realize that holes in cards could store information for weaving. He had cards made of thin wood, many measuring roughly 4 inches by 12 inches. Early cards had at most about 20 fairly large holes. They were strung together to form an endless loop. During the weaving process, the flat surface of a card kept hooked rods in position. When the tip of a spring-loaded rod reached a hole, the rod moved up through the hole, which lifted the warp threads. After the shuttle passed through, an assistant weaver reset the rods for the next card. Some intricate textile patterns used as many as 24,000 punched cards.

Jacquard's loom proved an instant success. It required fewer weavers, provided consistent quality with intricate patterns, and produced higher production rates. By 1812, France had 11,000 Jacquard looms. England had 100,000 in 1833. However, many weavers, who saw their livelihoods threatened, reacted against Jacquard looms. They destroyed looms and Jacquard personally suffered several violent attacks. In 1806, the French government stepped in and bought Jacquard's patents. Jacquard received a comfortable pension plus royalties for each loom built. He died in 1834, but using punched cards for information storage lasted far beyond his lifetime. Charles Babbage (1791-1871) used them for his 1834 analytical engine, the world's first computer.

Charles Babbage. Born near London in 1791 into a wealthy family, Charles Babbage did not have to work for a living. He received

Library of Congress

Joseph Jacquard, an early-1800s French weaver who used punched cards to store information.

Charles Babbage had a one-seventh-size brass model of his calculator made in 1832. Operated by a hand crank, it is the most celebrated icon from the earliest history of computing.

an excellent primary education and earned degrees from Cambridge University. The printed tables of trigonometric and other functions that navigators used to determine safe sea routes fascinated Babbage. Due to hand calculation and typesetting, the tables usually contained many errors. On occasions when a ship failed to return from a voyage, people often suspected that faulty navigation tables might have caused it to hit unexpected rocks and sink. Babbage wanted to create a device that would calculate mathematical values and automatically print them.

Babbage spent his own money paying talented machinists to make parts for the calculator that he designed. His calculator would use more than 4,000 gears, levers, cams, and linkages, all machined to exceedingly close tolerances. But Babbage had a difficult personality and alienated many of the machinists who worked for him, and his calculator was never completed. In 1832, he did assemble a one-seventh-size demonstration model, which still exists. Powered by a hand crank, this earliest automatic calculator operated flawlessly.

Babbage went on to consider a more general-purpose machine that would allow the user to determine mathematical tasks. He called his device an *analytical engine,* though modern terminology would describe it as a *digital computer.* Babbage decided to use the punched cards pioneered by Jacquard to store information for the device. He used one type of card to specify mathematical operations. Larger cards stored input data and controlled placement of answers in the machine's memory. During the 30 years Babbage worked on his computer, he also developed conditional branching, looping, and subroutines.

Construction of Babbage's computer turned out to be vastly more challenging than that of his unfinished calculator. The accuracy requirements for needed parts greatly strained 19th century technology. Babbage's difficulty in working with other people again got in the way of his invention, and the analytical engine was never built. Babbage had a crude demonstration model under construction at the time of his death in 1871.

Ada Lovelace (1815-1852) took a particular interest in Babbage's work with machine calculations. She brought to the project a proficiency in mathematics and she assumed public relations duties on Babbage's behalf. She also worked closely with Babbage on technical aspects of his projects, which led to her writing the world's first computer program.

Ada Lovelace. Born in London, Ada Lovelace's birth name was Augusta Ada Byron.

Charles Babbage had started the process of having this relatively crude iron model of his computer made when he died in 1871.

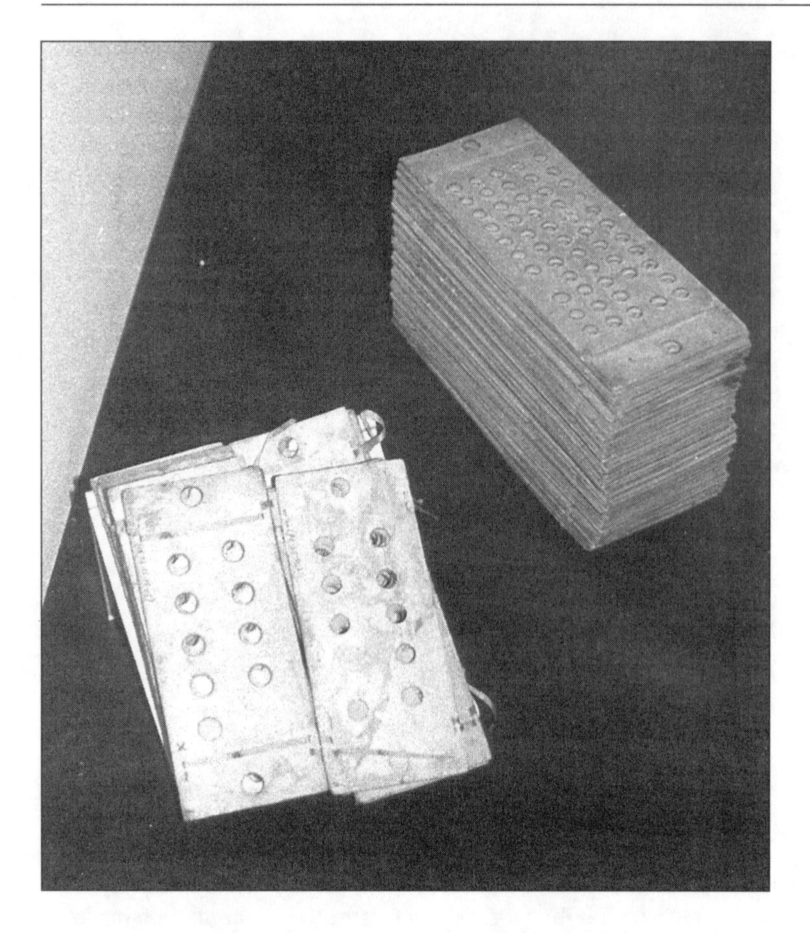

Ada Lovelace and Charles Babbage envisioned using punched cards for their 1834 computer. The larger cards at the right rear were for data and the smaller cards at the left front specified arithmetical operations. These are the only known Lovelace/Babbage cards still in existence.

She was the only child of British poet Lord (George Gordon) Byron and his wife Annabella Milbanke Byron. The couple separated when Lovelace was only a month old, and though she later corresponded with her father, she never met him.

Educated by tutors, Lovelace received special instruction in mathematics from a prominent mathematician, Augustus de Morgan (1806-1871). De Morgan, who contributed to the development of modern algebra, had a high opinion of Lovelace's abilities. She developed a close friendship with de Morgan's wife, Sophia.

After marrying a nobleman in 1835, Lovelace was called the Countess of Lovelace, or Ada Lovelace. In spite of social obligations and lifelong poor health, she continued to work on her mathematical skills through self-study. She regularly corresponded with Michael Faraday (1791-1867) and with scientific writer Mary Somerville (1780-1872). Somerville and Sophia Morgan introduced Lovelace to Babbage in 1833 and often accompanied her on visits to discuss calculations by machine.

In 1842, Babbage made technical presentations in Turin, Italy, about the computer he planned to build. He asked Lovelace to translate a French account of the lectures into English. In the course of translating the lectures, Lovelace added many of her own mathematical insights. The document she produces gives the modern world the details of Babbage's computer.

Babbage also asked Lovelace to write a computer program. She worked out a procedure for using Babbage's machine to calculate the Bernoulli numbers used in specialized calculus operations. Since the computer was never built, the program went untested during Lovelace's lifetime. But it would have worked. When used with modern computers, her program produces the expected values.

Babbage and Lovelace strongly wanted to see the computer built. But they would have needed a huge amount of money to produce its 200,000 close-tolerance parts and ensure their careful assembly. Unfortunately, the two devised a gambling system that they hoped would provide the required funds. But their system, which involved horse racing, failed. Both Babbage and Lovelace lost a great deal of money. Heavily in debt, Lovelace died at 36 years of age. She is buried next to her father, who also died at age 36, inside St. Mary Magdalene Church in the small town of Hucknall.

Pioneering Computers of the 20th Century

Early in the 20th century, some people mistakenly viewed calculators and comput-

Courtesy of Prof. Alan Hague

Ada Lovelace is buried inside St. Mary Magdalene Church in the small central England town of Hucknall. She is interred next to her father, British poet Lord Byron.

ers as being nearly identical. But a computer differed from a calculator in that it operated according to stored mathematical commands, which the user could change. Common 1940s methods for changing commands used punched paper tape, punched cards, rotary switches, and wire leads that had quick-disconnect terminals. The confusion about calculators and computers may have arisen around the use of *dedicated computers*, whose operation paralleled that of calculators. Dedicated computers performed a single complex task. One example was Britain's 1943 Colossus, which was only used for deciphering coded messages during World War II.

Before a general-purpose computer could be built, people had to establish its methodology and address the basic concepts of artificial intelligence. Babbage provided the groundwork for Britain's Alan Turing and America's John von Neumann, the two mathematicians who developed logic architecture for computers.

Howard Aiken built the first automatic general-purpose digital computer, the Mark I, and Grace Hopper took on the difficult task of programming the huge device. At about the same time in Germany, Konrad Zuse was working on his version. But Howard Aiken (1900-1973) got there first.

Early 1930s computers were analog devices that used graphical inputs and outputs. This one was used at the University of Manchester in England. An image of Alan Turing is barely visible on the back wall.

Howard Aiken. Born in Hoboken and raised in Indianapolis, Howard Aiken earned a Ph.D. from Harvard University. He stayed on at Harvard as a teacher and researcher, in part conducting research with automated calculators. In 1937, to identify numbers in binary code, Aiken devised a theoretical computer that used the on/off characteristic of electrical relays. Colleagues suggested that Aiken contact International Business Machines (IBM), a company that manufactured calculators, accounting machines, and other office equipment. IBM's president agreed to Aiken's proposal. The company ultimately paid two-thirds of Aiken's $500,000 project cost, and the U.S. government covered the rest.

Work began in 1939 on what IBM called the Automatic Sequence Controlled Calculator. Really a programmable computer, it was universally known as the Mark I. First operated in 1943, the computer was as long as a house, and measured eight feet in height and five feet in width. The Mark I's key hardware—3,000 clicking relays—made the computer sound like a room full of people knitting.

Operating the Mark I required adjusting up to 1,400 rotary switches and running four three-inch-wide punched paper tapes that held encoded programs. Three people took on this challenging task. One, Grace Hopper, went on to many later computer successes. The Mark I was mainly used to calculate trajectories for shells fired from large Navy guns. The first of its breed, it started America's leadership in the computer industry.

The 1943 Mark I computer used punched paper tape for program input. This is one of its four paper tape units. Notice the plugs for cables.

Library of Congress

This photo shows Grace Hopper sometime around 1961, when she served as the director of systems research at the Remington Rand Corporation.

Aiken went on to direct Harvard's computer facility, now called the Aiken Computational Laboratory. He retired to Florida in 1961 to work as a computer consultant and died at the age of 73. Over the years, he remained in contact with his long-time colleague Grace Hopper (1906-1992).

Grace Hopper. Born in New York City into a financially comfortable family, Grace Murray graduated from Vassar College with a mathematics degree. She later received advanced degrees from Yale University. She married Vincent Hopper in 1930 and started teaching at Vassar the following year. Hopper remained on Vassar's faculty until World War II broke out, when she joined the U.S. Navy.

Assigned to the Bureau of Computation at Harvard University, where Aiken's Mark I was located, Lieutenant Hopper joined officers Robert Campbell and Richard Bloch in 1943 as the world's first modern computer programmers. Her work so impressed Aiken that he asked her to write the Mark I's operating manual, the first publication of its type. Hopper continued working with Aiken on new computer versions until 1949. After leaving the navy, she worked as a senior mathematician for the company that later became the Remington Rand Corporation.

Hopper originated the idea of writing computer programs in English, rather than in complex machine code. Her work resulted in Flow-Matic, a language aimed at business applications. By the end of 1956, Hopper had company computers understanding 20 near-English statements. It took just one more step to get to the enormously popular COBOL, or Common Business Oriented Language. Hopper served on the organizing committee, and COBOL was based on her Flow-Matic language.

Rejoining the Navy in 1967, Hopper worked on standardizing payroll programs. Her 1985 promotion to Rear Admiral made her the first woman to hold the rank. The woman affectionately called "Amazing Grace" died at the age of 85. Navy destroyer USS Hopper was commissioned in her honor in 1997.

Americans were not the only ones to conduct early work with computers. Konrad Zuse (1910-1995) built a small experimental computer in Germany in 1935, unaware that Americans were at work on similar projects. With the world preparing for war, communication between the U.S. and Germany was strained.

Konrad Zuse. Born in Berlin, Konrad Zuse studied civil engineering in college. After graduating, he worked for the Henschel Aircraft Company. During his spare time, Zuse began building an experimental Z1 computer in his parents' kitchen. Made using Erector Set parts, it took three years to complete and was the size of a four foot cube. Zuse punched calculation instructions into 35 mm film and passed it through a program reader. His ingenious binary-number memory system used rod positions, one direction for on and the other for off.

The Henschel company supported Zuse's work on an improved Z2 computer. He was the only German permitted to develop computers during World War II. Germany's political and military leaders thought they would win the war long before a computer became operational. Like the Z1, Zuse's Z2 was not intended for production use. But his 1939 Z3 was. Zuse used 2,600 electromagnetic relays in a method that paralleled Aiken's Mark I.

When completed in 1941, the Z3 went immediately into service in the German aircraft industry. Its components included an operator's console, a program reader, and cabinets for the relays. It was first used for calculating airplane vibrations during the high stresses of battle. Like the Z1, the Z3 used instructions punched in 35 mm film. Operators reentered data for every new calculation and viewed a series of small lamps to read the binary number output. The Z1, Z2, and Z3 were all destroyed in a 1945 air raid.

After the war, Zuse formed a small computer company. He supervised the 1960 construction of a Z3 replica for the Deutsches Museum in Munich. Zuse died in Huenfeld at the age of 85.

Mid-20th Century Electronic Computers

Experimenters were investigating the theory behind electronic computer logic in 1937, and Britain's Alan Turing was among the first to consider the structure of the mathematics involved. John von Neumann extended Turing's work in America. He did more than anyone else to establish the architecture of electronic computers in the late 1940s.

Construction of various computers took place fast and furiously after World War II. The first significant large-scale electronic computer was the 1946 Electronic Numerical Integrator and Calculator, or ENIAC. In 1951, the Universal Automatic Computer, or UNIVAC, came along. The UNIVAC was the first commercially successful computer. But all of these computers required some type of program methodology. Turing first addressed that need.

The concept of the computer suggested a form of machine intelligence. No one had successfully constructed such a device. Even the Mark I used external programs in the form of punched paper tapes. Alan Turing (1912-1954) knew that developing machine intelligence would require methods unlike anything seen with externally controlled devices.

Alan Turing. Born in London, Alan Turing was the second of two sons. Because his father traveled extensively overseas with the British civil service, his parents placed the boys in foster homes in England. This proved an unhappy circumstance for Turing. Withdrawing into himself, he developed an interest in the operation of the human mind.

Turing turned to mathematics for answers and showed talent for the subject in secondary school. He graduated from King's College at Cambridge University in 1935 and stayed on as a tutor. In 1937, Turing described a theoretical computer in precise mathematical terms in a technical paper titled "On Computable Numbers." His machine could scan instructions encoded in its memory and make adjustments as required. Turing formalized the concept of the theoretical computer. The operation of his imaginary device paralleled the way a human mind thinks.

A shy, reserved person who enjoyed running and cycling, Turing served as a code breaker for the British government during World War II. He worked at a mansion north of London called Bletchley Park, where the bulk of such work was conducted. Turing's skills received little notice from the government, even though he played a crucial role in developing devices to decipher enemy radio transmissions.

After the war, Turing began work on designing the Automatic Computing Engine at the British National Physical Laboratory. A scaled-down version was built in 1950. Turing was among the first people to write articles on artificial intelligence, and computers that operated according to his methodology were called "Turing Machines." Suffering from severe depression, Turing took his own life at the age of 41. When *Time* magazine developed its list of the 100 Most Important People of the 20th Century, Turing turned up as the only computer pioneer

This 1960 replica of the Z3 computer by Konrad Zuse is at Munich's Deutsches Museum. The sloped section of the panel in front is for binary number inputs and the vertical panel is for showing the binary output with lights. The program tape reader is immediately at the right and the many vacuum tubes for binary switching are at the rear.

Smithsonian Institution

John Mauchly (right) and J. Presper Eckert (left) confer with an unidentified Army officer in the late 1940s. What appears to be the ENIAC computer is behind them.

on the list. Hungarian-American John von Neumann (1903-1957) extended Turing's work to its logical conclusion.

John von Neumann. The oldest of three children, John von Neumann was born into a well-to-do Budapest banking family. The "von" in his name indicated a minor rank of nobility of his father's family. A child prodigy, von Neumann could divide eight-digit numbers in his head at the age of 6. He also spoke ancient Greek with his father. He received degrees in engineering and mathematics from Zurich and Budapest universities. While still in college, he gained a reputation for his skill with advanced physics and mathematics.

When Princeton University invited von Neumann to its campus in 1930, he decided to immigrate to America. He became one of the original six mathematics professors at Princeton's Institute for Advanced Studies (IAS) and retained that position the rest of his life. Albert Einstein (1879-1955) was one of his colleagues.

Having taken American citizenship just before World War II, von Neumann worked on the Manhattan Project at Los Alamos, New Mexico. He was a major member of the group that designed the implosion-triggering unit for the first atomic bomb. That work gave von Neumann familiarity with elementary computers, and he expanded his efforts after the war.

Like Turing, von Neumann worked to develop a pattern that would allow electronic computers to use internally stored programs. The IAS asked him to develop a computer for its use and von Neumann incorporated some of his own ideas on computer logic. Its programmability would distinguish the IAS machine. Von Neumann insisted that computers must not simply be glorified calculators, with all their operations specified in advance. He said that they should be built as general-purpose logic machines that could execute a wide variety of programs. Such machines could react intelligently to the results of their calculations and be shifted easily from one task to another.

The IAS machine was not completed until 1952. Many advances in computer technology took place during the late 1940s, and the calculating power of von Neumann's machine was overshadowed by its competition. ENIAC, for example, was built in Philadelphia at the University of Pennsylvania, just across the Delaware River from Princeton.

Still, von Neumann's theoretical work proved pivotal to the design and production of greatly improved computers. Turing's efforts pointed him in the right direction, but von Neumann took computing concepts further than anyone else did. He is widely credited with devising the concept of the stored-program computer and its two essential components, a *memory* and a *control unit* for transferring information among different sections in the memory. Computer logic on all machines through the 1990s operated according to principles von Neumann developed, using what was commonly called "von Neumann architecture."

Von Neumann was often described as a cheerful man. He and his wife, Klari, had one daughter, who became a leading economist. Their home was the center of many social activities, with the von Neumanns serving as popular hosts. John von Neumann died of cancer at 53.

ENIAC. Early computer pioneers seemed to delight in developing pronounceable acronyms for naming their devices. ENIAC may have been the first. Sponsored by the U.S. Army to calculate the trajectories of artillery shells, it went operational in 1946.

Instead of using relays for binary switching, like the Mark I, ENIAC used 17,468 vacuum tubes. This first electronic computer took up more space than an average house. The walls of the large room that housed it were covered with electronic circuitry used for programming, with hundreds of rotary switches and patch cords. The task of programming often took two days or longer. The computer was generally programmed to solve problems in which one arrangement was used many times before another was needed.

A huge 30-ton machine, ENIAC drew 174 kilowatts of power. It was an externally programmed machine, like the Mark I, but it operated about 1,000 times faster. That significant advance put ENIAC in a class by itself. Nonetheless, its memory stored only the current equivalent of about three lines of text. Still, ENIAC laid the foundation for the modern computer industry. It demonstrated the possibility of performing high-speed digital computing with vacuum tube technology.

ENIAC was designed and built by J. Presper Eckert (1919-1995) and John Mauchly (1907-1980). The two men began a casual partnership in 1942, just after Eckert completed his Master's degree. Mauchly contributed his design talents, while Eckert had the engineering skill needed to build the computer. After ENIAC went operational, the partners saw a potential for its commercial prospects. And they held a letter from the University of Pennsylvania's president that gave them ENIAC's patent rights.

However, the university's new director of research viewed things differently and insisted that Mauchly and Eckert sign patent releases. They refused, left the university, and set themselves up as independent entrepreneurs. Others followed them, and the University of Pennsylvania never fully recovered from the effects of its ill-conceived decision.

The 1951 UNIVAC computer's processing unit was huge by modern standards. It used a separate console for input commands.

Neither Mauchly nor Eckert had business skills, but they managed well enough. They received a contract in 1946 from the U.S. Census Bureau to provide equipment to assist with compilation of the 1950 census. The device they proposed ultimately became the UNIVAC, or Universal Automatic Computer, the first commercially successful computer. Remington Rand would eventually build 46 UNIVACs. Remington Rand purchased Mauchly and Eckert's company, and brought the men into its Philadelphia organization.

UNIVAC. UNIVAC represented a huge technological advance over earlier designs. It was the first computer that had internal stored programs, or von Neumann architecture. UNIVAC had a main memory equivalent to about 1,000 words, but it also had a secondary memory stored on magnetic tape. It continued to use vacuum tubes, 5,400 of them. Its clock speed of 2.25 MHz was 18 times faster than ENIAC's top rate of 125 KHz. And, weighing only 8 tons and taking up only 352 square feet of floor area, it was considerably lighter and smaller. Remington Rand delivered the first UNIVAC to the Census Bureau in June 1951, which used the computer to tally part of the 1950 census. Depending on features, UNIVACs cost $1 million to $1.5 million each.

State governments, insurance companies, and the General Electric Corporation also purchased UNIVACs, for payroll calculations. Many business machine companies quickly established research and develop-

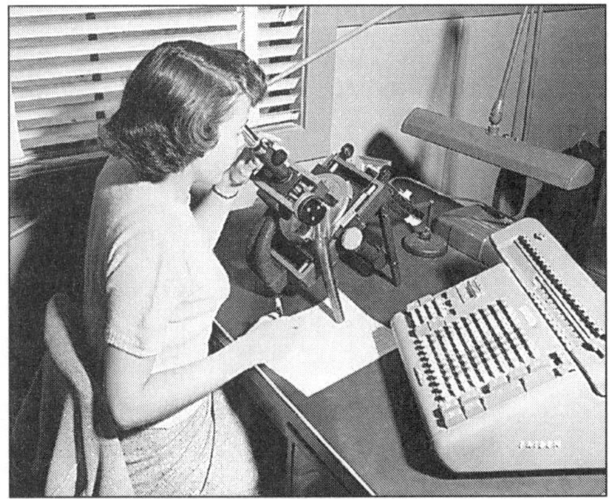

NASA

The woman in this photo is shown using a popular Friden electro-mechanical calculator at her desk. When this photograph was taken in 1952 at NASA's Ames Research Center, talented workers like her were often referred to as "computers."

Smithsonian Institution

John von Neumann stands beside the EDVAC computer in about 1952.

ment programs. These companies brought out their own style of computers and fierce competition in the industry began in earnest.

ENIAC and UNIVAC were not the only computers to use initialisms for names. Two other notable ones were:

• ABC—The John Atanasoff (1903-1995), Clifford Berry (1918-1963) Computer. Atanasoff worked as a professor at Iowa State University, where Berry studied at the graduate level. The two may have completed parts of a prototype computer seen by Mauchly in 1941. Atanasoff never applied for a patent nor made a complete computer. In a controversial 1973 decision, a judge invalidated Mauchly and Eckert's patent, ruling that Atanasoff had anticipated their ideas.

• EDVAC—The Electronic Discrete Variable Automatic Computer was intended to succeed ENIAC. Those working on its concepts included von Neumann, Mauchly, and Eckert. In late June 1945, von Neumann completed a lengthy document titled "First Draft Report on the EDVAC." It set forth the design of stored program computers. Built at the University of Pennsylvania, EDVAC went operational in 1952, too late to compete with others.

Computer Teams. As technology grew more complicated, ambitious innovators realized that they needed others to help them reach their goals. Such teamwork was not new. James Watt and Matthew Boulton worked together to construct steam engines in 18th century Britain. Joseph Brown and Lucian Sharpe built machine tools during America's industrial revolution in the 19th century. Charles Rolls and Frederick Royce assembled motor cars in the 20th century, and Walter and Olive Beech built airplanes. So it's not a great surprise that teamwork played an important role in computer technology.

William Hewlett and David Packard. William Hewlett (1913-2001) and David Packard (1912-1996) met as students at Stanford University. Hewlett had been born in Michigan and Packard in Colorado. The two engineering classmates became close friends. After graduation, Packard went to work for General Electric in Schenectady but returned to California two years later. Hewlett and Packard formed their partnership in 1938.

The Hewlett-Packard (HP) Corporation was one of the first in the Silicon Valley, just south of San Francisco. The men started it in a small garage in Palo Alto with an initial working capital of $538. HP's first major product was an audio oscillator, the best in the market. Walt Disney Studios purchased eight to test sound equipment for the 1940 animated motion picture *Fantasia*. From this successful start, the partners never looked back.

Hewlett and Packard built a global electronics, instrumentation, and computing company that was noted for its informal corporate culture. Hewlett was generally remembered for his engineering expertise and Packard for his business skills. Early on, they focused on manufacturing electronic test and measurement equipment. The company's first computer was the 1966 HP-2116. About the size of a file cabinet, it served as a controller for programmable instruments. Released in 1968, the HP-9100 was the first desktop scientific calculator. Then, Hewlett issued a company challenge to design a calculator that would fit in his shirt pocket. The result was the HP-35, the first pocket scientific calculator, released in 1972.

After they became very wealthy, Hewlett and Packard shared their good fortune in countless charitable ways. One of America's leading philanthropists, Hewlett served on the board of directors for hospitals and public policy organizations. At the time of his death at 88, the Hewlett Foundation had a net worth of $3.5 billion.

Packard served as U.S. Deputy Secretary of Defense during Richard Nixon's first administration. He served on national government councils and committees until he was 80 years old and left his entire estate to charity upon his death. The David and Lucile Packard Foundation supported community groups, hospitals, and youth agencies. Its net worth was $9.8 billion in 2000.

Stephen Jobs (1955-) credits William Hewlett with encouraging his youthful interest in electronics. While in junior high school, Jobs personally spoke with Hewlett by telephone and asked for parts for a computer he was working on. He got not only the necessary parts, but also a summer job at Hewlett-Packard. Establishing Apple Computer Corporation in 1976, Jobs and Steven Wozniak (1950-) made the first successful personal computers.

IBM's 704 computer was among the first to have the FORTRAN computer language provided with each installation. This photograph was taken in 1954 at IBM's computing center on Madison Avenue in New York City. Note the magnetic-tape storage cabinets at the right.

International Business Machines Corporation

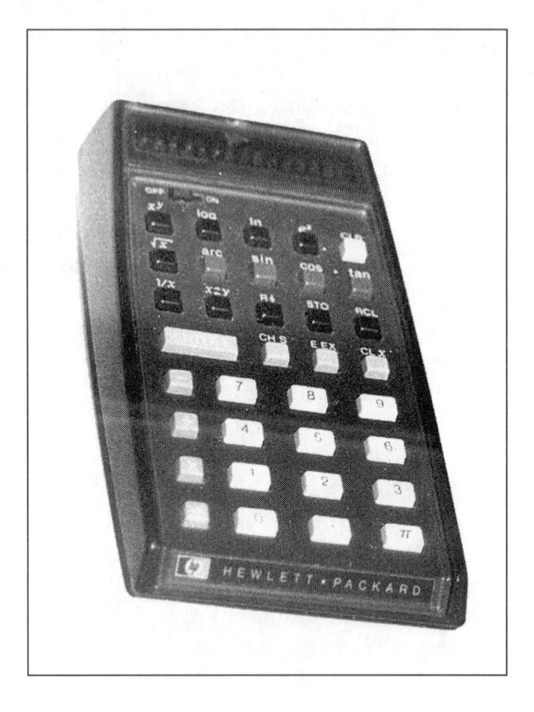

Hewlett-Packard's 1972 model 35 was the first scientific calculator small enough to fit in a pocket. The HP-35 was faster than the first large computers like ENIAC.

Stephen Jobs and Steven Wozniak. Both Stephen Jobs and Steven Wozniak were born in California, Jobs in San Francisco and Wozniak in nearby Sunnyvale. An orphan, Jobs was adopted by Paul and Clara Jobs. He showed an early interest in electronics and worked on various devices in his spare time. While in high school, Jobs joined the Explorers's Club, which Hewlett-Packard had established to encourage students interested in engineering. He got his summer job with HP when he was about 16 and met Wozniak there.

Wozniak had also developed an early interest in electronics. At the age of 13, he constructed a transistorized calculator that won a prize at the Bay Area Science Fair. Though he attended two universities, he graduated from neither. In 1971, while working at HP he met Jobs.

Jobs and Wozniak became close friends and both joined the Homebrew Computer Club, the first PC users group. With encouragement from other club members, they built a few early personal computers and developed some software for them. Though less talented with technology than Wozniak, Jobs had excellent marketing skills. They approached members of HP management with their ideas about producing personal computers, but the company showed little interest. The partners decided to set up their own company.

Jobs and Wozniak chose Apple for their company name because they wanted a simple word and Jobs had once spent a summer picking fruit in Oregon. They designed and assembled a prototype computer in 1976. To obtain money to market it, Jobs sold his Volkswagen and Wozniak sold his HP calculator. Having raised $1,300 and working out of Jobs's garage, they built the Apple I, which they sold for $666. This was the first single-board computer with a built-in video interface and on-board read-only memory. It had no keyboard and was targeted at people interested in electronics and computers.

The Apple I proved a popular product and Jobs and Wozniak easily found financing for their next computer. The Apple II, which came out in 1977, made it possible for the average person to purchase and use a PC. The first widely accepted PC, it cost $1,195 and had a 16K memory. Within three years, Apple had earned $139 million. The Apple II was the best-selling computer for five years in a row.

The Apple II proved particularly popular for educational use because of its low cost and readily available software. Jobs encouraged independent programmers to develop applications for the Apple II, which resulted in an early library of about 16,000 software programs. The computer became popular with business people after the world's first spreadsheet was written for it. Daniel Bricklin (1951-) and Robert Frankston (1950-) developed VisiCalc, offering it for sale in 1979 for $100. Many companies purchased Apple II computers just to use that particular program. All spreadsheet programs trace their heritage to VisiCalc.

With their company well established, Wozniak and Jobs went their separate ways. After suffering injuries in the 1980 crash

Even though it was not operated through use of a keyboard, the 1975 Altair 8800 is often considered the first commercially available personal computer. Bill Gates and Paul Allen established Microsoft Corporation to write software for it.

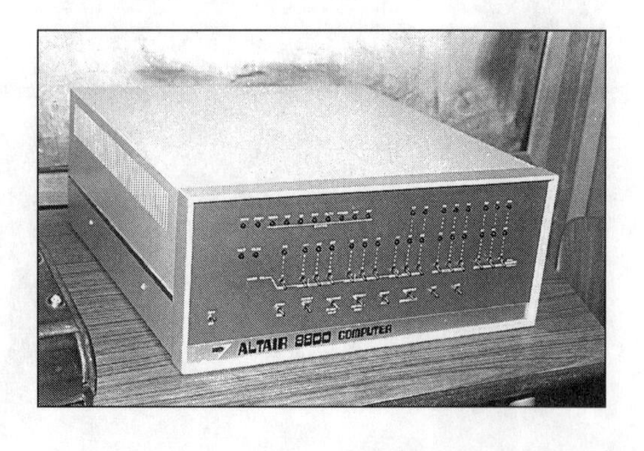

of his private airplane, Wozniak decided to return to college after recuperating. He earned a B.S. degree in computer science from the University of California at Berkeley in 1982. Jobs left Apple in 1985 to set up his own computer company, NeXT. In 1985, the two men were jointly awarded the National Medal of Technology, the highest American honor for technical achievement.

Jobs and Wozniak were ahead of the curve in recognizing the critical importance of computer software. Without programs, a computer is as useful as a television set that has neither an antenna nor a cable input signal. The operating system, which provides a method for telling the computer how to operate, is an essential piece of software. The most popular one was Microsoft Corporation's Disk Operating System, MS-DOS. Developed in 1981 by William Gates (1955-) and Paul Allen (1953-), it quickly became the industry standard. Gates and Allen founded the first personal computer software firm, Microsoft, in 1975.

William Gates and Paul Allen. William Gates and his two sisters were born into a comfortable upper-middle-class family in Seattle. He always ranked at the top of his class in elementary school, and his parents enrolled him in Lakeside Prep School partly because it had the resources to challenge his abilities.

In the 1960s, computers were expensive, cumbersome, and difficult to operate. In addition, high schools did not typically have access to them. But Lakeside struck deals with various corporations that would allow its students to use company computers installed at the school. The teenage Bill Gates met Paul Allen at Lakeside in 1968. They and a handful of others became so fascinated with the computers that they often skipped classes to work with them. Gates and Allen wrote programs, read computer books, and tried to figure out how the computers worked. They quickly mastered BASIC and FORTRAN computer languages.

Allen was born in Mercer Island, Washington, and he had a younger sister. He found his early work with computers so satisfying that he took a job writing programs for Honeywell International in Boston. Gates entered Boston's Harvard University in 1973 and the two friends continued their computer endeavors through after-hours work. Gates took computer classes and did programming at the Aiken Computational Laboratory. In 1975, after reading an advertisement for a microcomputer kit, Allen persuaded Gates that they should form a company to create software for the computer. Gates and Allen established the Microsoft Corporation and moved to Albuquerque, where the kits were produced. By 1978, Microsoft had $1 million in sales and had relocated to Seattle.

Stephen Jobs and Steven Wozniak put the prototype for their 1977 Apple II in a wooden case. A mannequin beside the computer gives an indication of its scale.

Microsoft went through some rough years but succeeded wildly after licensing its MS-DOS system to International Business Machines (IBM). Though IBM had originally focused on large mainframe computers, it decided to enter the personal computer market in 1981 with its Model 5150PC. Commonly called simply the IBM PC, the computer took the public by storm. Microsoft grew rapidly as the company continued producing software for the business and consumer markets. The company's first Windows version came out in 1987, and by 1993 at least a million copies sold every month. In 2001, Microsoft had over 40,000 employees in 60 countries. *Forbes* magazine rated Gates as the world's richest individual and Allen as third richest. (Investor Warren Buffett came in second.)

Allen was serving as Microsoft's head of research and new product development, when he experienced health problems and decided to leave daily work at Microsoft in 1983. A bachelor, he retained a seat on the company's board of directors and went on

The 1977 Apple II was the first commercially successful, keyboard-operated PC.

to spend his time with philanthropy, civic activities, and investments.

Gates remained with the company as chief executive officer and chairman of the board, and he wrote books about his professional experiences. He married Melinda French in 1994 and also involved himself more with philanthropy. Emphasizing health care and literacy, the Bill and Melinda Gates Fund had a worth of over $20 billion in 2000.

Birth of the Internet

Universities and research centers were among the first organizations to use computers on a regular basis. Some university and research center staff members envisioned the creation of a computer network that would permit different organizations to share information. In 1969, the U.S. Department of Defense agreed to fund a pilot project called the Advanced Research Project Agency, which resulted in the name ARPAnet. Early member organizations included Stanford University, the Massachusetts Institute of Technology (MIT), Harvard, and the Rand Corporation.

The ARPAnet concept was particularly notable because its organizational pioneers started to build a network at a time when computers were simply mathematical machines. Modems did not exist, different computers had different operating systems, and no standard existed for machine-to-machine communication. A group of researchers came together and worked for six months to establish a file transfer protocol (FTP). An early FTP, completed in 1972, allowed computers to communicate with each other. Within five years, the ARPAnet had 200 on-line members.

In 1973, Ray Tomlinson proposed using the @ sign in e-mail addresses. An MIT graduate, Tomlinson worked at the technical consulting firm of Bolt Beranek and Newman (BBN) in Cambridge, Massachusetts, and BBN contributed to the early stages of ARPAnet. Tomlinson sent several messages from one machine to another during test transmissions. Though he later said that he couldn't precisely remember the content of his very first message, he speculated that it was most likely "QWERTYIOP or something similar." By 1998, over 125 million people were using the @ sign in their email addresses.

ARPAnet users worked constantly to improve file transfer protocols. By the late 1970s, most people used one called TCP/IP, which stood for Transmission Control Protocol/Internet Protocol, a spin-off method developed by ARPAnet and the National Science Foundation. On January 1, 1983, ARPAnet changed over to TCP/IP, an event that marked the birthdate of the Internet.

Using the Internet in its early years was cumbersome and required precise keyboard inputs. An early browser, Mosaic, became available in 1993. Mosaic was the first user friendly, mouse-operated, clickable browser. Netscape Communicator came out a year later, and, by 1996, 80 percent of all Web surfers used it. By 2000, Microsoft's Internet Explorer ranked as the most popular browser on the Internet.

What started as a United States government project, has benefited everyone in the world. The Internet is universally available. It connects people with each other and with organizations. A phenomenal product of American know-how, it is unique in human history.

References

The Timetables of Technology edited by Bryan Bunch and Alexander Hellemans, Touchstone Publications, 1993.

"Father of the Computer Age" by I. Bernard Cohen, *American Heritage of Invention and Technology*,

Spring 1999.

Scientists and Inventors by Anthony Feldman and Peter Ford, Facts on File Publishers, 1979.

"Amazing Grace" by J.M. Fenster, *American Heritage of Invention and Technology*, Fall 1998.

The Computer from Pascal to von Neumann by Herman H. Goldstine, Princeton University Press, 1972.

"William Hewlett, A Pioneer of Silicon Valley, Dies at 87" by John Markoff, *New York Times,* 13 January 2001.

"How Von Neumann Showed the Way" by T. A. Heppenheimer, *American Heritage of Invention and Technology,* Fall 1990.

"Seventy Years of Engineering; Hewlett-Packard Company," *Machine Design*, 23 September 1999.

"UNIVAC in History as First Computer" by Reid Kanaley, *Lexington Herald Leader,* 29 April 2001, p. F2.

"Another First Computer" by Frederic D. Schwarz, *American Heritage of Invention and Technology,* Spring 1999.

Portraits in Silicon by Robert Slater, MIT Press, 1989.

Charles Babbage and His Calculating Engines by Doron Swade, Science Museum Publication, 1991.

Ada—Enchantress of Numbers narrated and edited by Betty Alexandra Toole, Strawberry Press, 1992.

Calculating Machines and Computers by Geoffrey Tweedale, Shire Publications, 1990.

CONNECTING CIVILIZATION

Biographies

Johann GUTENBERG

People communicate with words in several different ways. Visual information from the Internet is one example. Listening to the radio, telephone, or instructional lectures are others. Although possessing the advantage of immediacy, these media can't match the heritage or staying power of the printed page. Little printing today uses individually positioned pieces of metal type. Computerized techniques are now the preferred method. But using hand-set type is a hobby for some people. They make limited print runs for personal enjoyment. Like any hobbyist, these people have a greater interest in quality than quantity. They might print small books, hand bind them, and present them as gifts. Such beautiful publications often use very old presses and are sometimes considered to be an art form.

Books sold before the mid-1400s were costly and rare. Each was individually handwritten using quill pens or printed from hand-carved wooden pages. In either case, the books were too expensive for average people to buy. A high-quality handwritten Bible cost about 80 guilders in fifteenth-century Germany. For comparison, a stone house cost 80 to 100 guilders. The development of reusable metal printing type used with a printing press led to a drop in book prices and an increase in level of education. It started a revolution in mass communication. The first book produced using such methods was an impressive two-volume large-format Bible printed by Johann Gutenberg in 1456. Of the 200 copies that Gutenberg printed, 47 still exist.

Gutenberg was born in an industrial city on the Rhine River not far from Frankfurt. He was the youngest child in his family and may have had two siblings. He received training as a goldsmith or silversmith and earned his living at that trade. His parents were apparently well-to-do but lost their fortune and social status following a revolt by the lower-class citizens of Mainz. To escape the political upheaval, Gutenberg's family moved to Strasbourg, France, about 100 miles to the south, in 1411. The younger Gutenberg spent many years there and developed an interest in the possibility of mechanical printing. Goldsmiths marked their work with a metal punch and that activity may have fostered Gutenberg's interest in printing with replaceable, or *movable*, type.

Like other printers at the time, Gutenberg began by obtaining a block of wood as large as a page. He carved rows of letters that made words, inked the block, and pressed a sheet of paper onto to it. The process worked

Courtesy Deutsches Museum, München

Born:
1397 (?), in Mainz, Germany

Died:
February 3, 1468, in Mainz, Germany

The Gutenberg Bible at the Gutenberg Museum in Mainz is displayed under security glass in a darkened vault. Flash photography is not allowed. This is one volume of a two-volume Bible that was purchased in 1978 for about $2 million.

fairly well but was too labor intensive to be practical. He tried carving individual letters from wood. He personally carved many hundreds and combined the letters into words. Tightening them within a metal frame and pressing paper against the inked letters produced a readable image. However, that method was also labor intensive and the wooden letters wore out more quickly than Gutenberg had expected. That resulted in smudged words and unevenly printed letters.

Gutenberg returned to Mainz in 1448 and began work on a method that used interchangeable metal type. To purchase the necessary supplies, he borrowed 800 guilders from wealthy lawyer Johann Fust (1400 (?) -1466). That was a very large amount of money, since a master craftsman at the time typically earned 20 to 30 guilders a year. To make individual letters, Gutenberg engraved each character in relief on the end of a steel punch. He used a heavy hammer blow to strike the image into a piece of softer copper and formed a matrix. He then filed the copper matrix to fit in his two-part mold. The mold was an ingenious invention and the most important part of his printing system. Gutenberg poured in molten lead with a little tin and antimony to improve flowability and hardness. The lead alloy solidified within seconds, Gutenberg split the mold, and a cast letter fell out. The mold was adjustable for different letter widths, wider for a "w" and narrower for an "i". He could easily make as many identical metal letters as he wanted. Considering the alphabet, punctuation marks, and special characters, he needed about 150 different Latin characters for his Bible. On a good day, Gutenberg and an assistant could cast 600 lead characters. Each one had to be filed to a uniform height and stored in cases.

Gutenberg also developed an oil-based ink for use with metal type. The water-based ink used with quill pens tended to form into droplets and would not properly spread over the surface of the metal letters. Borrowing an idea from artists, Gutenberg mixed ground carbon with boiled linseed oil and produced a suitable ink. He also had to deal with hard-surfaced papers made for quill pens. He solved the problem by slightly moistening the sheets before pressing the inked type onto their surface.

Gutenberg came from a grape-growing region and adapted his printing press from one used to squeeze grapes for wine. Turning a large screw with a heavy wooden handle forced two horizontal plates together. The lower plate held the inked type in a clamped frame. The upper plate pressed paper against the type, forming an image on the paper. After printing several copies of one page, the type was removed and reset for another page. A photo on the next page shows a reconstruction of Gutenberg's printing press at the Gutenberg Museum in Mainz. His printing rate was about 30 pages per hour and the method remained essentially unchanged for more than 300 years. Gutenberg used six presses that probably cost him 40 guilders each.

As was common in his day, Gutenberg wanted his first major effort to have a religious connection. He and his assistant Peter Schoeffer (1425 (?)-1502) assumed the four-year task of setting all the type for a two-volume 1,286 page Bible. It was printed in Latin with 42 lines per column. It eventually required 3 million hand-set characters. Gutenberg purchased a high-quality handwritten Bible to use as a master. His printed book turned out beautifully. Although almost impossible for a nonscholar

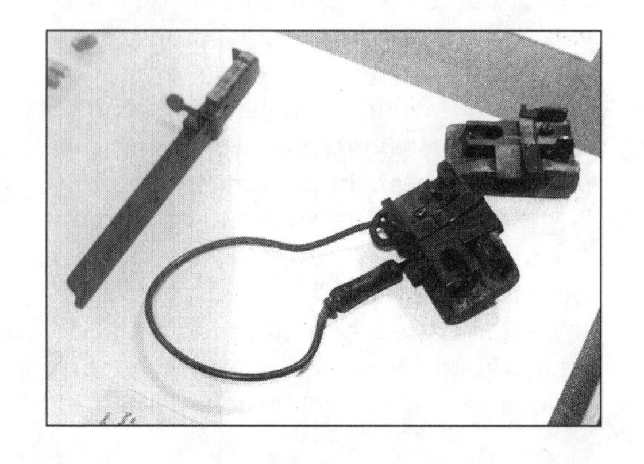

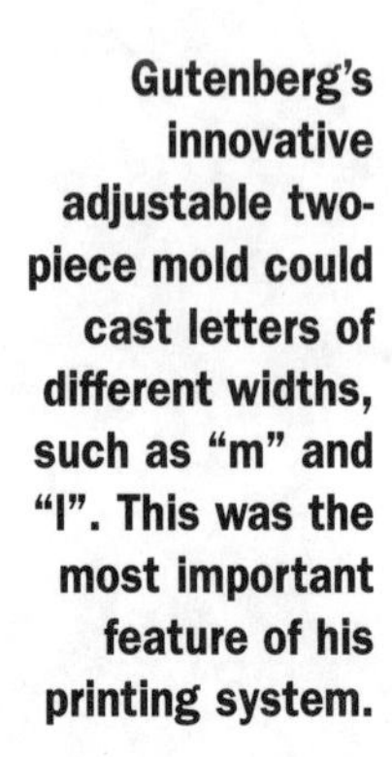

Gutenberg's innovative adjustable two-piece mold could cast letters of different widths, such as "m" and "I". This was the most important feature of his printing system.

This replica of Gutenberg's printing press and print shop are at the Gutenberg Museum in Mainz. Demonstrations are regularly given for visiting school groups. With little mechanical advantage to the screw on the press, it took a great deal of effort to press the inked type against the paper.

to read because it is written in Latin with an ancient type face, few modern books can rival its impressive appearance. Each sheet was elegantly type set, had decorative marks at the beginning of sentences, and was printed on the best Italian paper or parchment. Gutenberg printed each sheet so that the front and back were precisely aligned. He used tiny needles in the paper as positioning guides. Bookbinders put the sheets together for him.

The Gutenberg Bible was the first substantial publication to come from a printing press. One on display at the Gutenberg Museum was purchased in 1978 for more than $2 million. It is called the *Shuckburgh Bible*. Original copies in America are at the Library of Congress, Yale and Harvard Universities, the New York Public Library, and the Huntington Museum in California. It is hard to identify other books Gutenberg may have printed because he never put his name on his work.

The introduction of reusable type and a printing press provided a spark for other potential printers throughout the world. Within the next 45 years, about 15 million books had been printed. Gutenberg's innovation had changed human civilization forever. Books and magazines could be printed at low cost. That created an emphasis on literacy and led to the establishment of major universities and specialized fields of knowledge.

Gutenberg's indebtedness to Fust eventually amounted to more than 2,000 guilders, an incredible amount of money. As was his legal right, Fust foreclosed on the loan. Gutenberg could not come up with the repayment and Fust acquired all his tools and supplies. Gutenberg was almost reduced to poverty. Late in life, he received an adequate pension from the German government and died at about the age of 71.

For all his abilities as a professional printer, Gutenberg left no written material about his personal life. He has been a very difficult person for historians to research. About 30 documents from his lifetime mention his name, but most are legal papers. Gutenberg was continually plagued with financial problems of his own making. Only two documents make any mention of his printing profession, and those comment on it only in passing. It has been impossible to gain any insight into his personality or motivation. But there is little doubt about Gutenberg's technical successes. Secondary sources clearly verify his accomplishments. Although his abilities were little appreciated in his own time, he found a unique niche in history. He printed the only ancient book immediately known to much of the general public, the Gutenberg Bible.

References

The Gutenberg Bible by Martin Davies, British Library Press, 1996.

Stories of Great Craftsmen by S. H. Glenister, Books for Libraries Press, 1939 (reprinted 1970).

Printers and Printing by David Pottinger, Harvard University Press, 1941 (reprinted 1971).

Alessandro VOLTA

Born:
February 18, 1745, in Como, Italy

Died:
March 5, 1827, in Como, Italy

Electricity isn't as new a technology as some might think. The Greek scientist Thales (636 (?) - 546 (?) B.C.) made a basic electrical observation more than 2,500 years ago. He rubbed a piece of amber, fossilized tree resin, with a cloth and saw that it attracted small bits of straw. Thales was the first person to purposely generate static electricity. The word *electricity* is from the Latin *electrum*, which means "amber." In 1646, British physician Thomas Browne (1605-1682) was the first person to use the word.

Scientists in the 1700s built electrostatic generators to produce high voltages. The generators resembled a large plastic dinner plate, set vertically, and rotated with a crank. Metal braid rubbed against the plate and electricity collected on the surface of metal globes. It did not flow along a wire. That is why the energy was called *static electricity*. *Static* means "motionless." Bringing the globes close together produced a momentary spark.

Electrostatic generators were sometimes used for entertainment. Incompetent individuals occasionally generated voltages that seriously shocked some people. Many considered electricity only a laboratory curiosity until Alessandro Volta made the first battery in 1800. Volta's was the first device to produce electrical energy that could continuously flow down a wire. It was a stack of metal and cardboard disks called a *voltaic pile*. Benjamin Franklin (1706-1790) coined the word *battery* several years earlier. *Battery* refers to a group of items arranged together. Franklin connected several electrical storage devices called *Leyden jars*.

Pioneers of Electrical Communication by Rollo Appleyard, 1930

Volta was born near the Swiss border in northern Italy. His hometown was on a beautiful lake that attracted many European vacationers. His family was of local nobility. Though once financially comfortable, Volta's father had overspent the family fortune. They had fallen on hard times by the time Volta was born. He was the youngest of seven children. Other relatives from the extended family helped raise him.

Volta showed little early evidence of becoming a world-recognized scientist. He did not talk until he was four and his parents feared that he might have a mental deficiency. But he proved to be one of the brighter students in school. His father died when Volta was seven and relatives paid for his education through the age of 16.

He showed special ability with drama, foreign languages, and chemistry. Aside from Italian, Volta spoke English, French, and Latin, and he could read Dutch and Spanish. Volta was particularly impressed with the work of Joseph Priestly (1733-1804). Priestly was a British-American chemist who discovered oxygen. A two-volume book Priestly wrote, *History of Electricity,* inspired the young Volta to follow a technical career.

Volta's uncles wanted him to become a lawyer, a profession well represented on his mother's side of the family. But he chose to study electricity at the Royal Seminary in Como. After completing his studies sometime around 1763, Volta remained in Como for several years and began to establish a technical reputation. His early experiments included rubbing different metals with cloth and evaluating their electrical charges. He also described an improved electrostatic generator. Volta published his results over the years and they helped him gain a position in 1775 as a physics teacher at the high school in Como.

For some time, Volta had been working on a clever electrical device that he finally completed in 1775. He called it an *electrophorus*. The device used two metal plates, each about six inches in diameter. One was coated with ebonite and remained on a table. Ebonite is a dark hard rubber similar to plastic. The other plate was uncoated and had an insulating handle extending from its middle. Rubbing the ebonite with a cloth produced a negative charge. It attracted the positive charges to one side of the metal plate when the two plates were brought close together. A small wire drained negative charges from the other side and left the metal plate with extra positive charges. Continuing the process over and over built up a substantial positive charge on the metal plate. The amount of charge could get quite high and the invention brought Volta publicity and fame. The electrophorus has no practical use in the modern world other than to demonstrate static electricity in educational settings.

The University at Pavia, near Milan, offered Volta a professorship in 1779. He accepted and stayed for the next 40 years.

This unlabeled electrostatic generator is a 19th-century Wimshurst machine, named for James Wimshurst (1832-1903). Electrical charge from the spinning disk is stored in the two slender Leyden jars on either side. When enough charge builds up, a spark jumps between the two metal globes. The photograph in the background shows a modern demonstration unit safely dissipating electricity into the air though the girl's hair.

To keep up with the latest developments, Volta traveled extensively. One lengthy trip in 1781 took him to Switzerland, France, Germany, Belgium, Holland, and England. In recognition of his contributions to the fields of electricity and chemistry, the British Royal Society awarded Volta its coveted Copley Medal in 1794. The medal was a precursor to the Nobel Prize.

Volta worked primarily with static electricity until he heard of observations made by Luigi Galvani (1737-1798) in Bologna. In 1791, Galvani announced the existence of animal electricity. When two probes made from different metals touched a frog leg, the muscle twitched. Galvani speculated that electricity was stored in the leg. His theory had a large following, but not everyone agreed with it. Volta strongly disagreed and began his own investigations.

Volta was convinced that the presence of two different metals was all that mattered. His experiments resulted in his battery in 1800. It consisted of a series of cups partly filled with vinegar. Volta connected them with U-shaped strips of metal. One end of each metal strip was copper and the other end was zinc. Volta put his fingers in the two end cups and felt a slight tingle. The tingle decreased as the number of cups was reduced. It stopped when a metal strip was removed. Volta knew he had produced electricity.

To make the battery more compact, Volta used round disks of copper, zinc, and cardboard soaked in vinegar and stacked one on top of the other. The battery generated about one volt for each set of disks. Volta referred to "30, 40, 60, or more pieces of copper, or rather silver, applied each to a piece of tin, or zinc, which is much better. . . ." The *voltaic pile* was the first source of continuous current electricity and its effect was immediate. Within a few weeks, William Nicholson (1753-1815) and Anthony Carlisle (1768-1840) in England used it to separate chemical compounds. Andre Ampere (1775-1836) in France and Georg Ohm (1789-1854) in Germany added to the development of current electricity.

Northern Italy took part in military actions that involved the French emperor Napoleon I (1769-1821). Napoleon established the Cisalpine Republic in 1797, which included such cities as Bologna, Milan, and Pavia. Napoleon was essentially the ruler of northern Italy. He was fascinated by Volta's battery and encouraged the National Institute of France to ask him to lecture and give demonstrations in Paris. It took Volta 26 days to make the journey in September 1801. He stayed for four months. It was the beginning of a wave of rewards for the modest and talented university professor.

Volta received France's Legion of Honor and was made a count of the French Empire. Napoleon provided Volta with a lifetime pension of 4,000 lire per year and, in 1809, his annual salary increased to 24,000 lire. For comparison, at the time, a small house cost about 14,000 lire. Volta had the income of a wealthy man during the last 20 years of his life.

Volta was a pleasant person who made friends easily. Perhaps because of his noble heritage, Volta had the bearing of a gentleman. He did not marry until he was 49 and had a most happy life with the former Teresa Peregrini. She was an expert in agriculture and botany and spoke French and German. The couple had three children and were devoted parents. Volta's eldest son became mayor of Como and wrote many studies of his father's life. Volta retired to his family's country estate in 1819 and died at 82.

Volta was in his mid-50s when he invented the battery. After that, he came to a complete stop. He took no part in applying his discovery to the new fields it opened up. During the remaining 27 years of his life, he showed none of his earlier creativity. Before 1800, Volta had published 80 articles or books, many on significant topics. But after 1800, he produced only six unimportant works. Nobody knows why. The 1881 International Electrical Congress established the *volt* as the standard unit of electrical pressure.

References

Allesandro Volta and the Electric Battery by Bern Dibner, Franklin Watts Publishers, 1964.

Electrical Engineers and Workers by P. W. Kingsford, Edward Arnold Publishers, 1969.

Electricity by Steve Parker, Dorling Kindersley Publishing for the Science Museum, 1992.

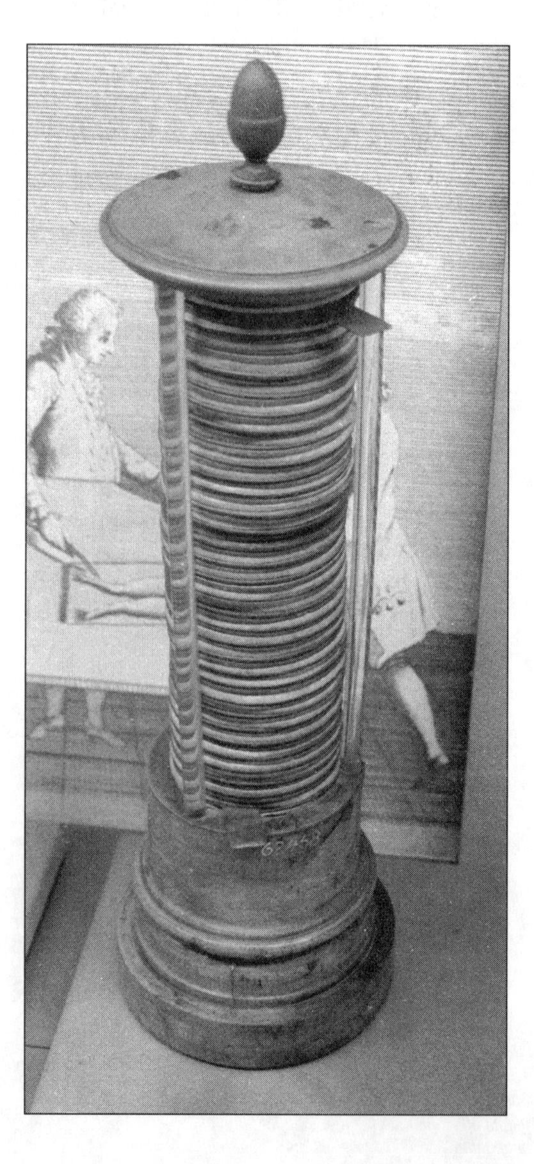

This voltaic pile has a label that describes it as a column. Volta obtained positive (+) electricity with a wire connected to the copper disk at the top and negative (-) electricity from the zinc disk on the bottom.

Andre AMPERE

Born:
January 22, 1775, in Lyon, France

Died:
June 10, 1836, in Marseilles, France

People who work with new ideas in technology, often fall into one of two categories. Those who work with pencil and paper are *theoretical* investigators. Those who work with hardware are *experimental* investigators. As a practical matter, most scientists, engineers, and technologists perform both types of investigation. When inventors sketch an idea, for example, they act as theoreticians. When they construct their prototypes, they act as experimentalists. But many technologists have tended to focus on one approach. Josiah Willard Gibbs (1839-1903) was America's only native-born theoretical scientist with a worldwide reputation. He did most of his work in physical chemistry with pencil and paper. By contrast, Ferdinand Zeppelin (1838-1917) of Germany was an experimentalist. No one before him had ever constructed a successful rigid-frame airship. His first was built mostly by trial and error, and was an experimental activity.

The Science Museum/Science & Society Picture Library

The field of electricity has provided plenty of theoreticians as well as plenty of experimentalists. Italy's Allesandro Volta (1745-1827) made the first battery and Germany's Georg Ohm (1789-1854) constructed the first electrical circuits. Both were primarily experimentalists. The unit of electrical pressure, the *volt*, is named for Volta, and the unit of resistance, the *ohm*, is named for Ohm. The *amp*, or *ampere*, is a unit of electrical flow named for the theoretician Andre Ampere. Ampere's best work came through his use of advanced mathematics to analyze the relationship between electricity and magnetism. He established the field known as *electrodynamics*, which studies the effects caused by electricity flowing in a wire.

Ampere was born in south-central France, the second of three children in a prosperous business family that dealt with textiles. His father taught him at home and Ampere quickly developed a huge appetite for printed materials. He read the entire French encyclopedia and showed an aptitude for mathematics. Most of the city library's books on mathematics were in Latin. So at the age of 12, Ampere taught himself Latin so he could read the books. He had an uncommon talent for languages and a lifelong interest in the possibility of a universal language.

Ampere was a teenager during the French Revolution (1789-1799) and probably understood little about it. A revolt against the monarchy of Louis XVI (1754-1793) and Marie Antoinette (1755-1793), it started in the north of France. Ampere's father supported the monarchy and was involved with the city of Lyon's attempt to maintain the old government. Revolutionaries captured the city, convicted Ampere's father of treason, and executed him in 1793. The younger Ampere was devastated and did not speak for a year. It was the first of an number of personal tragedies Ampere would experience.

Ampere met a young lady, Jule Carron, who brought him out of his depression. The couple married in 1799. They had one son. Ampere at first taught mathematics at a high school in Lyon. Then, he found a better-paying job at a similar school in Bourg, about 40 miles away. The distance was too great for him to return home every evening. He spent the week in a small apartment and went home on weekends. The forced separation from his family gave him time to work on publications about the statistics of gambling. The theory of probability got its start with the mathematics associated with games of chance, like dice and cards. Ampere expanded his work into advanced topics like probability calculus and began to gain a reputation in mathematics.

After several years of illness, Ampere's wife died in 1803 of an unknown ailment. Ampere again endured the loss of a loved one. He began to look around for a job that would take him away from Lyon and its unhappy memories. Based partly on the quality of his published works, Ampere found a position as a mathematics instructor at the Polytechnic [Technical] School of Paris, about 250 miles north of Lyon.

Unhappy in a large and unfamiliar city, he eagerly sought companionship with the Potot family, which befriended him. He married Jeanne Potot in 1806. But Ampere's new father-in-law swindled him out of some money and his marriage proved unhappy. He and his wife had a daughter before they dissolved their marriage. From that time, Ampere raised both his children in Paris with the help of his mother and an aunt.

Ampere's professional life followed a few twists and turns over the next several years. In 1808, he became inspector general for the newly formed university system. He became a member of the mathematics department of the Imperial Institute in 1814. Ampere's job required frequent travel and he chose to name his scientific findings after the places where he first thought of them. Ampere had a "theory of Avignon," a "demonstration of Genoble," and a "theorem of Montpellier." He was a member of many professional organizations. The French Academy of Sciences was among the most important. He was named assistant professor of astronomy at the University of Paris in 1820. That same year, he heard about a remarkable electricity and magnetism observation made by Hans Oersted (1777-1851) in Denmark.

Current electricity was a new idea and technologists often argued over any connection it might have with magnetism. Many people thought there was no relation between the two. Oersted was professor of physics at the Polytechnic Institute in Copenhagen. He conducted an experiment in which he placed a magnetic compass needle directly below a wire. When electricity flowed through the wire, the needle deflected slightly. This easily repeated experiment suggested a direct link between electricity and magnetism. Oersted's minor observation put Ampere on the path to worldwide recognition.

Within two weeks of hearing of Oersted's discovery, Ampere was working out the math. He began a series of six weekly reports to the Academy of Sciences. He showed that electrical flow in a circle acts like a bar magnet and that an iron bar placed inside the coil becomes magnetized. Ampere called this a *solenoid*, from the Greek word

These three unlabeled Leyden jars probably date from around 1800. They stored electric charge and were precursors to modern capacitors. They were metal-foiled glass jars first made in 1745 at the University of Leyden in the Netherlands.

This early, unlabeled, moving coil milliammeter probably dates from about 1850. The manufacturer was Baird and Tatlock.

solen, which means "tube." The word is still used for electromagnetic relays.

Ampere proved mathematically that the force between two current-carrying wires is a function of the electrical flow, wire lengths, and distance between the wires. Electricity was too new for technologists to be certain of the flow direction. But Ampere showed that the direction of flow between two nearby wires influenced whether they were attracted or repelled. He called this Ampere's rule. Flows in the same direction attracted two parallel wires and Ampere built a simple apparatus to demonstrate the effect. He discussed magnetic forces around wires carrying electricity. All of this showed how clearly Ampere understood current electricity. His was the first work on electrodynamics.

Ampere was the first person to apply advanced mathematics to electricity and magnetism. In 1823, he stated a theory that magnetism came from the action of tiny electrical charges in the wire. In spite of Ampere's reputation, his colleagues were skeptical. Ampere was trying to describe an effect that was similar to saying that electricity is a flow of electrons. The electron was not discovered until 1897 by Joseph Thomson (1856-1940) in England.

Ampere's successful professional life contrasted with his personal life. His son, Jean-Jacques, had an involvement with a wealthy widow twice his age. He became one of many in her entourage. Ampere's daughter, Albine, married an alcoholic army officer and had a difficult life. Though he lived in Paris, Ampere died in Marseilles at the age of 61 while visiting a university. The 1881 International Electrical Congress established the *amp* as the standard unit of electrical flow rate.

Electricity is one technical area whose development was shared by people from many countries. Volta was Italian, Ohm was German, and Ampere was French. The unit of inductance, the *henry*, was named for the American Joseph Henry (1797-1878). The unit of magnetic flux, the *tesla*, was named for Nikola Tesla (1856-1943) a Croatian-American. The unit of capacitance, the *farad*, was named for Michael Faraday (1791-1867) of England. The unit of energy, the *joule*, was named for James Joule (1818-1889), also from England.

The unit of power, the *watt*, came from James Watt (1736-1819) of Scotland. The unit of electrical charge, the *coulomb*, was named for Charles Coulomb (1736-1806) of France. The unit of frequency, the *hertz*, was named for Heinrich Hertz (1857-1894) of Germany. No field of human endeavor has been more international than electricity.

References

Andre-Marie Ampere by James R. Hoffman, Blackwell Publishers, 1995.

Pioneers of Electrical Communication by Rollo Appleyard, Macmillan Publishers, 1930.

Electricity by Steve Parker, Dorling Kindersley Publishing for the Science Museum, 1992.

Robert, Richard and *Robert* III HOE

Robert Hoe:
Born: October 29, 1784, in Hoes, England

Died: January 4, 1833, in New York, New York

Richard March Hoe:
Born: September 12, 1812, in New York, New York

Died: June 7, 1886, in Florence, Italy

Robert Hoe III:
Born: March 10, 1839, in New York, New York

Died: September 22, 1909, in London, England

Between 1898 and 1927, about 60,000 Stanley Steamer automobiles were manufactured in Newton, Massachusetts. The company was started by twins Francis Stanley (1849-1918) and Freelan Stanley (1849-1940). It was America's first successful automobile company. The Stanleys production rate was adequate for the early 20th century, though it would be quite inadequate today. Approximately 10 million motor vehicles are currently made in America every year. As the population increases and lifestyles change, consumers often want more or better products. Newspapers and other printed materials were not immune from those demands.

Johann Gutenberg (1397-1468) made his printing press in the 1450s by adapting a grape press used in wine making. Turning a large screw with a broomstick-sized handle forced two horizontal plates together. Between the plates lay a sheet of paper and a clamped form holding inked type. The pressure printed an image on the paper. Gutenberg's press had a printing rate of about 30 pages per hour. Centuries later, Benjamin Franklin used a similar screw-type press to print his *Pennsylvania Gazette* in 1730. In England, Charles Stanhope (1753-1816) made the first all-metal flat-bed press in 1800. It was stiffer and made sharper impressions than the wooden variety. But all hand-operated, flat-bed presses had low printing rates. Robert Hoe's introduction of his rotary cylinder press in 1830 made low printing rates a thing of the past. The high-speed cylinder press became the workhorse of all newspapers during the 1800s. It allowed newspapers to expand their readership and to provide more pages of information per issue.

National Portrait Gallery, Smithsonian Institution

Robert Hoe III (1839-1909)

Robert Hoe was the first member of a printing family that greatly affected American life. He was born into a prosperous farming family in central England. His father apprenticed him to a carpenter when he was about 15. Robert heard about advancement opportunities in America and he emigrated in 1803, while still a teenager. Hoe made his living as a carpenter and soon married Rachel Smith. He changed professions to help his wife's brothers, Peter and Matthew Smith, manufacture screw-type printing presses. The type that Gutenberg and Franklin used, they sold for about $400.

The Smith brothers both died in 1823 and Hoe inherited the business. He went looking for a new printing product to manufacture. He heard of the flat-bed cylinder press invented in Germany by Friedrich Koenig (1774-1833). In 1814, *The Times* of London was the first newspaper produced on the new steam-powered press. A clamped frame held the inked type on the flat bed of the printing press. A steam engine rotated a cylinder that rolled sheets of paper against the type. An operator hand fed each sheet onto the cylinder at a printing rate of 1,200 sheets per hour. The previous rate with high-speed hand-operated flat-bed presses was 200 sheets per hour. But the Koenig press cost $4,000 without the steam engine, a high price. Hoe sent one of his best workers to London to see the press in action.

Steam power was not widely available in America. Based on his worker's reports, Hoe made some changes to Koenig's design. He offered his own hand-cranked cylinder press for sale. The first machine was purchased in 1830 by the *Temperance Recorder* of Albany, New York. The printing rate was about 400 pages per hour. Hoe was starting to introduce steam-powered presses when his health began to fail. He transferred the business to his son Richard, just before he died prematurely at age 48.

Richard Hoe, his family's oldest son, was educated in the public schools of New York City. His father brought him into the factory at 15 and gave him various jobs to help him learn the business. He was the most technically proficient member of the family and expanded on his father's work. Richard was the first person to successfully attach printing type to the cylinder. The cylinder rotated continuously, which greatly increased production rate. His patented hand-cranked single-cylinder flat-bed press printed 2,000 pages per hour. He followed it with a double-cylinder press in 1837. All hand-operated presses required strong men to turn the cranks. They had to be frequently relieved. The increase in business encouraged Richard Hoe to construct more buildings for manufacturing his presses. His company soon had four acres under roof.

Richard Hoe made a huge leap forward with his 1846 patented steam-engine-powered press. He eliminated the flat bed entirely. Four impression cylinders contacted one cylinder carrying type. At each impression cylinder, a worker fed individual sheets of paper into the press. One rotation of the type cylinder printed four sheets. Its production rate was 10,000 pages per hour. The workers had to be fast because at top speed, the cylinders drew paper at the rate of about one sheet per second from each person. The first new press was installed in 1847 by the Philadelphia's *Public Ledger* newspaper. Later Hoe presses had as many as 10 impression cylinders and sold for $25,000.

Also called a type-revolving press, Hoe's cylinder printing press overshadowed all others. It soon replaced those used by newspapers throughout America, Europe, and Australia. With higher production rates, newspapers became more profitable. The R. M. Hoe Company was the leading manufacturer in the world. In the late 1860s, Hoe opened a branch factory in London that soon employed 600 persons. William Bullock (1813-1867) found a way to print from a continuous roll of paper in 1865. He called his innovation a *web feed,* after bolts of woven fabric, which were called webs. The method proved particularly useful in print-

This Hoe newspaper press was built around 1900 from designs developed in the 1870s. It printed and folded 48,000 four-page newspapers every hour.

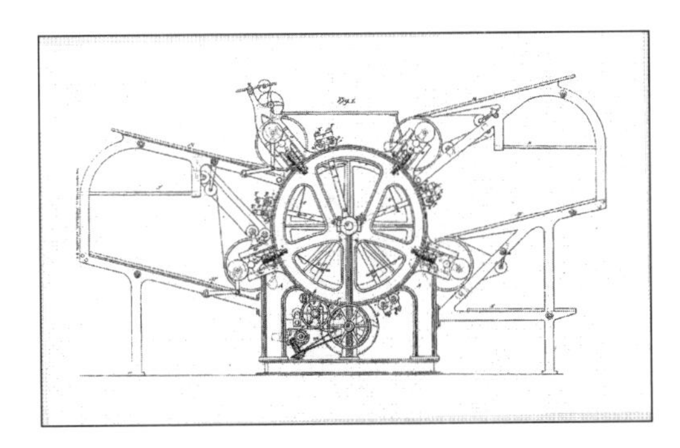

This is a drawing from Richard Hoe's patent for a four-impression cylinder rotary press. The cylinder at the bottom center carries the type. Workers stand on scaffolds and feed individual sheets of paper along the four ramps.

This is a model of Robert Hoe's early hand-cranked rotary press. The inked type is locked into the flat bed. The paper is fed from the plate at the top and carried along the outside of the cylinder. Cranking a handle rotates the cylinder and moves the type under the paper, to form an impression.

ing newspapers. The Hoe family quickly arranged to use Bullock's invention with their presses.

The company was an enlightened organization and had an early employee benefits program. Workers received a free lunch, partly subsidized medical care, a cooperative company store where they could purchase food at reduced prices, and free night school classes. Richard Hoe married twice and had three children. He watched over every detail of his factory until he suffered from overwork. He went overseas to rest and recover. He died at the age of 75 while visiting Italy. His nephew Robert Hoe III, grandson of the man who started the company, was the next family member to take over.

Robert Hoe III did not have the technical insight of his predecessors. However, he had the ability to select people who could carry out the improvements he thought most important. Hoe directed his energies toward meeting customer's requests for improved production rates. In 1891, his company constructed a cylinder press of 16,000 parts for the *New York Herald*. Using a continuous roll of paper, the web press printed, cut, folded, and counted 72,000 eight-page newspapers per hour. Such presses cost between $40,000 and $80,000.

One of the founders of New York's Metropolitan Museum of Art, Robert Hoe III appears at the start of this chapter in a 1891 drypoint engraving. He married Olivia James in 1863 and they had five children. Robert Hoe III wrote and published *A Short History of the Printing Press* in 1902. An avid collector of rare books, he had over 20,000 when he died unexpectedly at the age of 70.

Production printing has seen more changes than most other technologies. Flatbed printing gave way to cylinder printing using hand-set type. Hoe presses were supreme throughout the world for about 25 years after 1847. Cylinder printing easily merged with the Linotype method that used melted type to form a page. It was invented by German-American Ottmar Merganthaler (1854-1899). In July 1886, a portion of the *New York Tribune* was set with an experimental Linotype typesetter.

Hoe's presses were very important to the spread of news throughout an expanding 19th-century America. In a free society, literate and informed citizens depended on inexpensive newspapers. Between 1830 and 1860, the number of American newspapers increased from 863 to 3,725. Although there are fewer modern newspapers, each of the top four print more than one million copies a day. Their computerized techniques are far removed from Gutenberg's original printing press, but they provide more flexibility. All printing methods had the same objective, to get information to people on a printed page. Each accomplished that objective as best it could during its time in history.

References

The Story of Printing by Irving B. Simon, Harvey House Publishers, 1965.

American Journalism by Frank Luther Mott, Macmillan Publishers, 1962.

A History of American Manufactures from 1608 to 1860 by Leander Bishop, Samson Low, Son and Co. Publishers, 1868.

The Printers by Leonard Everett Fisher, Franklin Watts Publishers, 1965.

The first Hoe cylinder press was used to print a short-lived specialty publication in Albany, New York, called the *Temperance Recorder.*

Georg OHM

Born:
March 16, 1787, in Erlangen, Germany

Died:
July 7, 1854, in Munich, Germany

It would have been impossible for early 19th-century technologists to predict the future importance of electricity. They saw it only in the form of bolts of lightning and harmless shocks that resulted from scuffing shoes over a rug. The word *electricity* comes from the Latin *electrum*, which means "amber." Amber is fossilized tree sap similar to modern plastic. Rubbing it with fabric produced static electricity that attracted small scraps of paper and lifted animal fur. In the early 19th century, most people considered electricity to be more of a curiosity than a potentially useful form of power.

It became an important emerging technology after the invention of the storage battery in 1799. Alessandro Volta's (1745-1827) battery was a small stack of dissimilar metals and cardboard. Volta was not the only person investigating electricity. Others around the world were adding their bits and pieces. Michael Faraday (1791-1867) built the first electric motor in England in 1821. Andre Ampere (1775-1836) mathematically analyzed electricity in France in 1827. Joseph Henry (1797-1878) worked with electromagnetism in America in 1830. Another international player gave us a simple electrical formula that many people remember all their lives. Georg Ohm experimented with electrical currents and included his results in an 1827 German book. We now call his formula Ohm's law: voltage equals current times resistance. It is the most widely used relationship governing the behavior of electrical circuits. In formula form, it is $E = I*R$. E stands for electromotive force, I for intensity, and R for resistance.

Georg Ohm was born in south-central Germany, not far from Nuremberg. (His first

The Science Museum/Science & Society Picture Library

name is pronounced "gay-org.") Ohm's parents had seven children, though only three survived to adulthood: Georg, younger brother Martin, and sister Elisabeth. They were raised by their father after their mother died in 1797. Georg and Martin enjoyed technology and supported each other's activities. Martin became a respected mathematician at the Military College in Berlin. Their father was a master locksmith who liked science. He encouraged his children to read and taught the brothers the use of precision tools. He showed them how to fabricate the high-quality parts needed to construct a reliable lock. That skill would become the foundation of Ohm's future success.

Ohm attended the local high school, hoping to learn about the technical concepts his

The resistance value of Ohm's resistors varied with length and diameter. He wrapped them with insulating fabric, usually silk, and then twisted them around a nail or wooden peg. This resistor was made and used by Ohm in January 1826.

father introduced. But his education stressed classical instruction such as art, literature, and Latin. He grew up at a time in German history when educators felt that speculation was the only way to answer important technical questions. They had not yet accepted the idea that answers should come through experimentation and data evaluation. After high school, Ohm attended the university in his hometown, but he could not discipline himself. After he wasted time on social pursuits for a year and a half, his father decided that he was misusing the family's limited resources. Ohm's father insisted that he leave school. Ohm moved to rural Switzerland and started earning a living. He was a mathematics teacher and tutor for several years before returning to the University of Erlangen. This time he was better prepared for his studies and earned a Ph.D. in physics in 1811. He specialized in the subjects of mechanics and light.

Germany was then involved in the 1806-1814 Napoleonic Wars with France, which depressed civilian employment opportunities. Erlangen had a population of just 8,000 but had to shelter more than 33,000 soldiers. The possibility of a civil war was on everybody's mind. It was a difficult time for any young person looking for a job. Ohm wanted to teach at a university, but he decided to stay at home and work as a tutor for a short time. He finally found a full-time position teaching mathematics and physics at a high school in Bamberg. He stayed there for three years, working on his first book during his free time. It was a geometry book with a 13-word title that came out in 1817.

Although not outstanding, Ohm's textbook proved adequate. More important, its publication helped him obtain a similar position in a better setting. He taught at a religious high school in Cologne. The school had enthusiastic students and Ohm could use its laboratory. His first investigations were into the new concept of electromagnetism. It was 1825 before Ohm realized that research, followed by publication, might be the key to obtaining a university teaching position. Technical research and publication was a new practice just beginning to take place in Germany.

Focusing on how resistance influenced electrical current flow, Ohm was particularly interested in the conductivity of metals and their behavior in electrical circuits. But the science and practice of electrical measurement simply did not exist. Electrical technology was not well established and Ohm had to make his own meters and resistors. His resistors were of copper wire that he had learned to make while studying under his father. They were of remarkably consistent quality. Ohm laboriously wrapped the bare wire with insulating silk thread. Some wires were as thin as 0.025 inches in diameter and some as long as 75 feet. Each length was wrapped around a nail or wooden peg. Ohm's early crude ammeter was a compass needle hanging on a thread. His zinc-copper battery was approximately the shape of a cube, six inches on a side. Brine-soaked cardboard separated the metal plates.

Ohm conducted tests for several years and was the first to suggest an analogy between electrical flow and liquid flow in a pipe. His initial conclusions were disappointing because he could not repeat them. Ohm determined that the high internal resistance of the battery was influencing his readings. He was the first to make such an observation. When he felt confident enough, he published his conclusions. His 1827 book was titled *The Galvanic Circuit Mathematically Treated* and included what we now call Ohm's law. Ohm wrote, "The magnitude of the current (amperes) in a galvanic circuit is directly proportional to the sum of all tensions (volts) and inversely to the total reduced length (ohms) of the circuit." He used the term *total reduced length* for resistance because his resistors were wires of varying lengths.

Ohm's conclusions drew little attention from the technical community. Others could not duplicate his results. In part, they lacked the skills needed to make accurate instruments and test specimens. They were also held back by the early-19th-century German belief that experimentation and mathematics were irrelevant to understanding natural phenomena. Influenced by German philosopher Georg Hegel (1770-1831), many thought Ohm's formula too simplistic. They treated him with indifference and even hostility. Ohm was crushed. He felt he had done everything necessary to advance the field of electrical technology, but others misread his results.

Although Bavaria is now a state of Germany, it was a separate kingdom in the mid-1800s. Bavarian King Ludwig I had far-sighted technical advisors. He gave Ohm a professorship at the Polytechnic School of Nuremberg in 1833. This was the first positive acknowledgment Ohm received concerning his electrical discoveries. He was also appointed to the position of State Inspector of Scientific Education where he could oversee necessary educational reforms. The British used his book to guide many of their electrical experiments. They were so impressed with Ohm that in 1841 they awarded him the coveted Copley Medal, a precursor to the Nobel Prize. The photograph at the beginning of this chapter shows Ohm with the medal around his neck. German scientists saw their past errors and lobbied to offer Ohm a professorship at the University of Munich. He accepted this high honor in 1849 and became department head in 1852. Ohm died in 1854 while working on an optics experiment. He had understandably turned his back on electrical research.

Ohm was of average height, sturdy, and physically strong. He often wore a long, dark-blue coat with large pockets for his electrical equipment. He was an almost tireless worker who found comfort in his professional work. He never married. Ohm was an excellent teacher who approached problems with students as if he did not yet know the answer. He had a pleasant voice and an excellent sense of humor. He never let his professional disappointments affect his relationship with students or supportive colleagues. Although Ohm was displeased that powerful leaders worked against him, he never gave up on technology. And he achieved world recognition in his own lifetime. Few people have been as courageous as Ohm. Fewer still have both a law of nature and a unit named after them. In 1881, the International Electrical Congress in Paris, established the *ohm* as the basic unit of electrical resistance.

References

Pioneers of Electrical Communication by Rollo Appleyard, Books for Libraries Press, 1968.

The Communications Miracle by John Bray, Plenum Press, 1995.

Famous Names in Engineering by James Carvill, Butterworth and Co Publishers, 1981.

Louis DAGUERRE

Born: November 18, 1787, in Cormeilles, France

Died: July 10, 1851, in Petit-Bry-sur-Marne, France

It would be hard to imagine modern life without pictures. Colorful static images from newspapers, magazines, billboards, personal photographs, and the Internet almost demand our attention. Dynamic moving images in motion pictures, on commercial television, and in instructional videotapes appear everywhere. Imagery informs, entertains, and inspires.

19th-century technologists knew the power of the picture and looked for easier ways to capture life on paper, glass, or metal. The earliest technique was the *camera obscura*. The name comes from two Latin words. *Camera* means "chamber" and *obscurus* meant "dark." It was a box with a lens at one end and a 45° mirror at the other. Aimed at a building or rural scene, the mirror focused the reflected image onto a piece of ground glass on top. A person could then place a piece of thin paper onto the glass and sketch or paint the subject, using the reflected image as a guide.

Another image-recording technique was named after Etienne Silhouette (1709-1767), a French minister of finance. Although it is uncertain whether Silhouette originated the idea of forming an outlined profile of a person, his name describes such an image. Another Frenchman, Joseph Niepce (1765-1833), produced the first permanent photograph in 1822. But his technique required exposures of up to eight hours. Louis Daguerre introduced a more practical method in the 1830s. Requiring exposures of only a minute or less, it was fast enough to photograph people. His daguerreotypes caused a worldwide sensation.

Daguerre was born about nine miles northwest of Paris and grew up during a difficult period in his country's history. The French Revolution (1789-1799) affected the country's social programs and Daguerre received little formal education. His father, a minor government official, apprenticed Daguerre to a draftsman in his early teens. Daguerre had an aptitude for sketching and wanted to study painting. His father relented in 1804 and Daguerre went to Paris. He was fortunate in locating an apprenticeship with the chief stage designer at the Paris Opera House. He stayed at the opera house for three years and then went to work for Pierre Prevost, a painter well known for his panoramas.

Courtesy Deutsches Museum, München

Part of Daguerre's responsibility was to help paint large cycloramas that had become

This 1839 daguerreotype camera has a peep hole to help the photographer compose the image.

Prevost's specialty. In an era before photography and motion pictures, the viewer sat in the center of a large cylindrical painting of a single expansive subject. The paintings required great attention to accuracy of scale and perspective. Daguerre's experience with drafting served him well and he stayed at the Prevost Studios for nine years. He married Louise Smith in 1810 and chose to strike out on his own in 1816.

Daguerre made a living by painting scenery for many of the best-known theaters in Paris. The prestigious Academie Royale de Musique contracted with him on several occasions. He was one of its chief designers for two years. Daguerre perfected a clever scene-painting technique that allowed the background to appear to change, such as from summer to autumn. One of his best-remembered efforts was a mountain slope. Carefully turning lights on and off gave the set the appearance of a landslide occurring in the background. This was the most exciting optical illusion before motion pictures.

Such dramatic results began with Daguerre painting a scene onto a large piece of semi-translucent cotton fabric. A carefully applied thin coat of paint was almost transparent. To show summer changing to autumn, Daguerre would complete two paintings and place one in front of the other. Sometimes he had three in a row. A turpentine wash on the fabric made it even more transparent. By illuminating various sections of the painting from different angles and through different filters, Daguerre obtained a variety of exciting motion effects. The huge fabric sheets were as wide and high as an opera stage. Daguerre's work achieved an astonishing degree of three-dimensional realism and he received many forms of recognition. But none of his early work remains. Because of the combustible nature of the materials, a theater fire in 1839 destroyed everything.

In Daguerre's era, a person's financial condition was all important. He had no social status on which to build and felt obliged to use his talents and whatever else he could muster. Driven by a desire for wealth and publicity, Daguerre saw photography as a way to reach that end. He learned of Niepce's work in 1826 and read what he could on the topic. He knew that long exposure time was a limiting factor. Daguerre had no photographic experience and arbitrarily thought he could shorten exposure time to take pictures of people. He wrote a letter to Niepce suggesting a partnership. Daguerre misrepresented himself by saying he had some photographic successes when he actually had none. Niepce had received little financial reward from his innovation. He felt that the younger Daguerre was an energetic and optimistic person who could improve his process. The two signed an agreement in 1829.

Although Daguerre worked diligently on the project, he did not make any notable progress until after Niepce died in 1833. Sometime around 1837 and entirely by accident, Daguerre made two critical discoveries. He found that highly polished iodized plates could be used as a surface to capture images and that mercury vapor would develop the image. The mercury technique came about because Daguerre unknowingly broke a mercury-filled thermometer in a cabinet where he stored his plates. Today we would say the vapor fogged the plates.

Daguerreotype cameras were little more than light-tight boxes. The back of this one slid to allow focusing on a ground glass before the sensitized plate was put in place.

Daguerreotypes were commonly used for portraits during the mid-1800s. They were kept behind glass and often in velvet cases like this one because they scratched easily.

The effect gave Daguerre a useful clue.

Daguerreotype photography was the first method by which people could have their pictures taken. The exposure time was around one minute or less in bright sunlight. Lengthy by modern standards, it explains why so many early photographs show people with almost no expression. They had to sit perfectly still for a long time.

There were different ways to generate a daguerreotype. Here is one:

- Polish a silver-plated piece of copper.
- In a lighttight box, coat it with iodine vapor for 30 seconds.
- Using a camera, expose the plate in bright sunlight for 30 seconds.
- In a lighttight box, develop with vapors from heated mercury for 1 to 3 minutes.
- Fix the image in hyposulfite and water for 2 to 3 minutes.
- Protect the image by pouring on a gilding solution of 1 gram gold-chloride and 1/2 liter of water.

Since there was no negative, daguerreotypes could not be duplicated. Each photograph resulted in only one print. They were typically made in six standard sizes. The largest was 6-1/2 by 8-1/2 inches, and the smallest was 1-5/8 by 2- 1/8 inches. The gilding solution gave some measure of protection to the highly polished image. The gold coating and the silver plating give daguerreotypes their metallic appearance.

Like Daguerre, telegraph inventor Samuel Morse (1791-1872) started life as an artist, primarily as a portrait painter. Before becoming wealthy with his 1840 invention, Morse met Daguerre while traveling in Europe. He learned how to take daguerreotypes and opened one of America's first photographic studios in New York City in 1838.

Soon after he made it practical, Daguerre turned over his photographic rights to the French government. They had offered a lifetime pension of 6,000 francs per year, equivalent to about $1,200 and a very comfortable amount. Daguerre retired to a country estate in 1840 and let others improve his process. He returned to painting for personal enjoyment and died at the age of 63.

The daguerreotype dominated photography for more than 10 years. But it was a blind alley that was put to rest by William Talbot's (1800-1877) invention of negative-positive photography. Daguerreotypes were fragile and could neither be duplicated nor enlarged. But deadend designs have often helped point inventors to a better direction. Elias Howe's 1846 sewing machine was also a blind alley. Its cumbersome horizontal needle could make about a dozen stitches before the fabric had to be repositioned. Thomas Edison's 1877 phonograph used cylinders instead of more practical flat records. The Wright brothers 1903 airplane had rear-mounted pusher propellers with the flight control surfaces in front.

Beautiful daguerreotypes have had unusual staying power. Museum displays make them look as if they were exposed only yesterday. Photographer Matthew Brady (1823-1896) had studios in New York City and Washington, D.C. During the Civil War, he took countless poignant, troubling, and historic images. Many were daguerreotypes.

References

Daguerre by Beaumont Newhall, Winter House Publishers, 1971.

The History of Photography by Helmut Gernsheim, Thames and Hudson Publishers, 1969.

American Handbook of the Daguerreotype by S.D. Humphrey, Arno Press, 1973.

Samuel MORSE

It is not unusual for some people to start on a different career path before discovering technology. Even so, it seems unlikely that a talented painter should invent the first electrical communication device. It is even more unlikely that that person would be nearly 50 years old—most successful inventors start at half that age. The technologist referred to here, Samuel Finley Breese Morse, patented his telegraph in 1840.

Morse was born to a minister and his wife, both of whom appreciated the value of a proper education. Morse attended local schools and an academy for gifted youngsters. He received excellent instruction that prepared him well for studies at Yale University. There, Morse majored in art but also took classes in electricity, chemistry, and mathematics. Several of his teachers had the best technical minds in the country, but Morse never showed an interest in science or technology as a career. After graduation, his parents financed a trip to London that allowed him to study under the leading painters of the time.

The War of 1812, between America and England, broke out the year after his arrival, and Morse stayed overseas for four years. His travel was not hampered by the war, and he met many excellent artists. He returned to Boston in 1815, where he made his living as a portrait painter. The talented Morse once did a portrait of his friend Eli Whitney, which is now in the collection of Yale University's Art Gallery. New York City Hall and the New York Public Library also currently own and display his paintings. Morse helped found the National Academy of Design in 1826 and served as its president for 20 years.

A series of personal tragedies rained down on Morse during the 1820s. One of his children died while he was living in Charleston, South Carolina. Shortly after returning to New York, Morse's wife died, as did his father and mother. He left for Europe to seek inner peace and fresh ideas for his painting career. Morse spent most of the next three years in France and Italy. He met Louis Daguerre and was the first American to see a daguerreotype. He made plans to open a photographic studio on his return to the United States.

Smithsonian Institution Photo No. 10-799

Samuel Morse during the last year of his life

Born:
April 27, 1791, in Charlestown, Massachusetts

Died:
April 2, 1872, in New York, New York

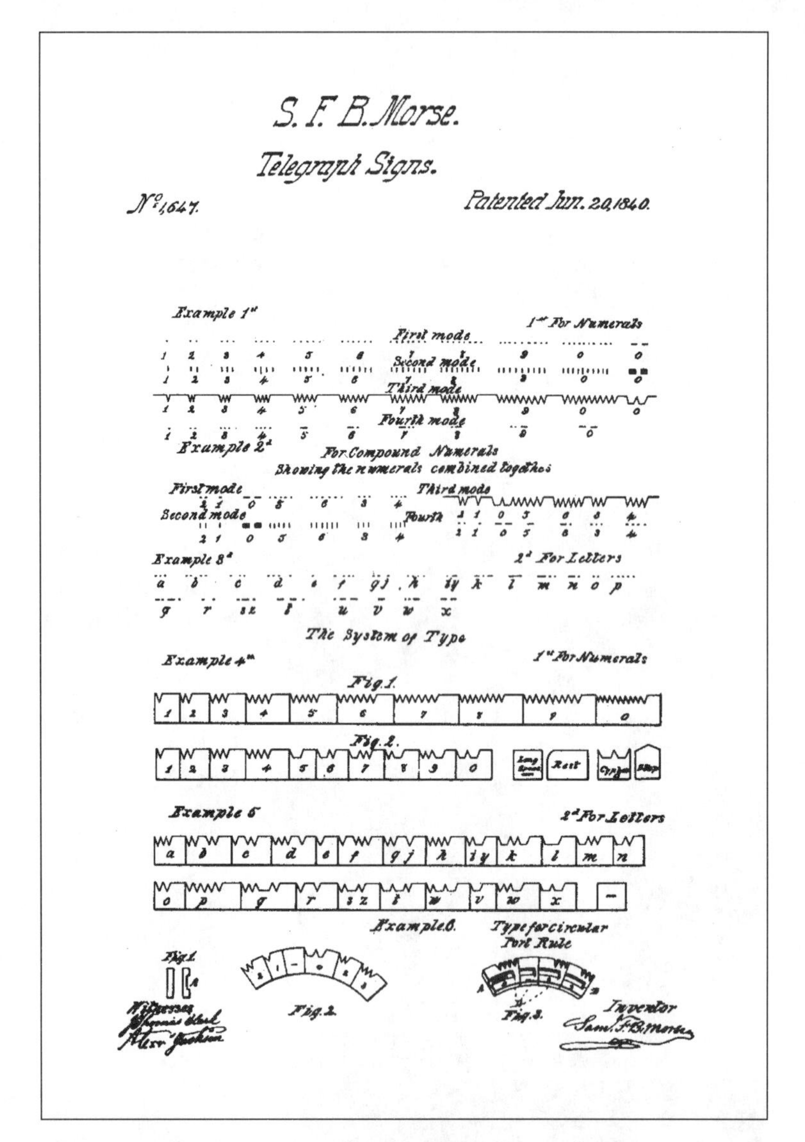

Morse's telegraph patent included his first attempt at a Morse code.

Morse sailed back in 1832. One of the other passengers on the voyage was Charles Thomas Jackson, a Boston physician. Jackson brought up the subject of electricity and its ability to travel almost instantaneously over any length of wire. The idea of applying this concept to the rapid transfer of information fascinated Morse, and he engaged Jackson in lengthy conversation. Later that evening, Morse started sketching a crude wood-framed telegraph. He used the little knowledge of electricity that he retained from his college days in this diversion during the crossing.

Immediately after landing in America, Morse sought work to support himself and three young children. He taught painting and sculpture at the newly opened New York University and functioned as the first professor of art in America. One of his students was Mathew Brady, the famous Civil War photographer. Morse also made and sold daguerreotypes, and he opened one of the first photographic portrait studios. He worked on the telegraph in his spare time. He redoubled his inventive efforts in 1837 after he heard that Europeans were making great strides in telegraphy. In England, Charles Wheatstone was developing a method in which needles pointed to letters on a dial. About that time, Congress turned down Morse's bid to provide a painting for the Capitol rotunda. Taking this as a final failure to gain recognition as a great artist, he decided to concentrate his efforts on the telegraph.

Morse soon discovered that he lacked adequate knowledge of electricity. One of his early electromagnets did not work simply because he used uninsulated wire. Since the electromagnet was basic to the telegraph, Morse approached Joseph Henry, secretary of the Smithsonian Institution. Henry had invented the electromagnet, and he freely gave all the help he could. Morse spent seven years trying different designs while selling partnerships and a few paintings to pay the bills. At one time, his net worth had dropped to 37 cents.

The receiver Morse patented in 1840 did not resemble the standard telegraph receiver common a few years later. His original was a printing instrument. Morse used a small wooden frame to support a pendulum with a pencil on the end. One wooden crosspiece near the bottom had a roll of paper tape and a clockwork mechanism that pulled the paper under the pendulum. Another crosspiece had an electromagnet near the center of the iron pendulum. The transmitter was a key that closed an electrical switch, and it had some similarity to modern keys.

Morse's transmission code was a series of dots and dashes. Briefly closing the transmitter key caused the electromagnet to move the pendulum a very short distance. That motion produced a "dot." Closing the key longer made a "dash." As the paper moved under the pencil-tipped pendulum, the pencil drew an image resembling a small mountain peak for each dot and a small plateau for each dash. The pencil lines formed a distinct pattern for each letter that Morse had devised. In the first version of the Morse Code, for example, the letter "T" had two dashes and a dot. "P" had five dots. Morse's receiver was a cumbersome affair, but it worked. By 1845, trained telegraphers listened to the dot-dash pattern from an electromagnet in a sounding box and wrote messages directly.

Morse gave demonstrations of his new equipment in Newark, New York, and Philadelphia. He took it to Washington in 1842 to demonstrate for government leaders. From the beginning of his experiments, Morse never saw his telegraph as simply a convenient method of communication. He thought it would be used only for messages of the utmost importance and hoped the government would buy his patent rights. Members of Congress showed no interest in buying, but they did approve a $30,000 grant for installing a 41-mile trial telegraph line between Washington and Baltimore.

Morse accepted a suggestion to use a single wire strung on glass insulators mounted on trees and poles. Using the earth as a ground eliminated the necessity for a two-wire system. The young Ezra Cornell completed the installation work. He would soon make a fortune in the telegraph business and provide money to establish Cornell University. The telegraph's terminal points were the Supreme Court chambers in Washington and the Baltimore and Ohio Railroad depot on Pratt Street in Baltimore. On May 24, 1844, from Washington, Morse tapped out the message, "WHAT HATH GOD WROUGHT?" It took one minute to send the four words to a partner, Alfred Vail, in Baltimore. The message was chosen by Annie Ellsworth. She was the young daughter of Henry Ellsworth, U.S. Commissioner of Patents and a Yale classmate of Morse's. Although this was not the first message ever sent by telegraph, it was the first official intercity message. The telegraph became an immediate success.

With the exception of Florida, every state east of the Mississippi River had telegraph service just four years later. London went on line in 1866. Morse lived to see all those early major developments in telegraphy. American colleges, foreign governments, and professional societies showered him with awards. Morse's years of poverty had ended, and he spent his remaining days as a wealthy man on his 200-acre estate on the Hudson River. The photograph on the first page of this profile is a silver print taken during the last year of his life.

References

Those Inventive Americans, National Geographic Society, 1971.

American Science and Invention by Mitchell Wilson, Bonanza Books, 1960.

"What Hath God Wrought?" by Maury Klein, in *American Heritage of Invention and Technology*, Spring 1993.

Dictionary of American Biography, Charles Scribner's Sons Publishers, 1932; with supplemental updates.

National Cyclopedia of American Biography, James T. White & Co. Publishers, 1891; with supplemental updates.

McGraw-Hill Encyclopedia of Biography, McGraw-Hill, 1973.

Asimov's Biographical Encyclopedia of Science and Technology, by Isaac Asimov, Doubleday & Co. Publishers, 1964.

Charles BABBAGE

Born: December 26, 1791, in Teignmouth, England

Died: October 18, 1871, in London, England

When technologists of the mid-twenty-first century look back at 20th-century inventions, it is impossible to predict which will emerge as the single most important. But, without doubt, the digital computer will be in the running. These electronic workhorses have a very wide range of uses. They control factory machine tool operations, monitor hospital patients, and, under the command of pilots, fly commercial airplanes. They function on a more personal level by helping individuals write letters and connect to the world through the global Internet system. They also entertain people with video games. The digital computer seems to have endless potential.

But the digital computer was not a 20th-century invention. It was born in England, before the reign of Queen Victoria. All modern computer pioneers credit Charles Babbage as being the first in their field. He designed all the essential parts of a digital computer in 1834. Working in an era before the use of electricity, he planned to operate his gear-driven device with a steam engine.

Babbage came from a wealthy banking family and never had to work for his living. Both he and his sister, Mary Anne, received excellent primary education at boarding schools. Their parents' only surviving children, they shared a close lifelong relationship. Self-educated in mathematics from an early age, Charles Babbage knew more than his college instructors at Cambridge University. After graduating in 1814, he married Georgiana Whitmore. In 1815, the couple moved to London, where they lived for the rest of their lives.

Babbage's father had helped finance business ventures that included transportation by ships. Navigators of the era used printed navigation tables to determine safe sea routes. The tables were both calculated and typeset by hand, which produced some errors. It was estimated the tables may have contained thousands of undiscovered errors. When, on occasion, a ship did not return from a voyage, it was common to assume that incorrect navigation tables might have caused it to hit rocks and sink. Babbage grew fascinated with the possibility of making a machine that could calculate and print navigation tables, which would eliminate errors in calculation.

The Science Museum/Science & Society Picture Library

In the early 1820s, Babbage started working on what he called a *difference engine*. This was actually a calculator that worked by adding and subtracting numbers over and over again. A *difference* is the amount by which one number varies from another. The word *engine* simply meant a device or machine. Babbage spent most of the next 10 years designing his difference engine. His plans called for 25,000 carefully made parts. The brass machine would be 6 feet tall, 6 feet long, and almost 3 feet deep. Thousands of gears, levers, cams, and linkages had to be made to exceedingly close tolerances. Babbage had his own machine tools and used them regularly. He even had a forge. But there was no way he could have made the necessary high-quality parts for his difference engine. Such a large and demanding project required the services of talented machinists. The best available worked for Joseph Clement (1779-1844) in Newington.

Babbage contracted with Clement, using his own money to finance the work. He soon saw that it would cost far more than he had expected. Babbage approached the directors of the Civil Contingencies Fund of the British government for a grant to continue the project. At the time, government funding of technical research was a new idea, and many politicians disapproved of Babbage's request. Other opinions prevailed, however, and the money became available in 1823.

Clement's machine shop worked on Babbage's calculator for four years. In later years, Joseph Whitworth (1803-1887), the person who standardized threaded fasteners, would proudly relate that he personally made many of the critical parts. Unfortunately, Babbage's difficult personality upset the gifted machinists with whom he worked. It has been speculated that the unexpected deaths of his wife, father, and two children in 1827 had a devastating and permanent effect on Babbage's temperment. For whatever reason, his calculator was never completed. He had spent about £6,000 of his own money and £17,000 of government money on the project.

To show some progress to British officials, he had a one-seventh size demonstration model assembled in 1832. It was 72 by 59 by 61 centimeters, and it had 2,000 parts. Powered by a hand crank, it operated faultlessly and was the earliest automatic calculator. For the first time, mathematical logic was a basic part of a mechanism. The beautifully made brass assemblage of gears, levers, cams, and rods has survived the years and is on display at London's Science Museum. It remains the finest example of 19th-century precision machining and is the most celebrated icon from the earliest history of computing.

Babbage's difference engine processed numbers the only way it could, by adding or subtracting them in a particular sequence. After designing it, Babbage thought about a more general purpose machine—one that would allow the user to determine all the mathematical tasks. He called this an *analytical engine*. Today it would be called a *digital computer*. The five processing sections Babbage developed in 1834 are the same as those used by modern computers:

- input device—punched cards,
- arithmetic unit—the calculating section,
- programmable unit—the part that determines the calculating sequences,
- memory—the gear positions, and
- output mechanism—printer, plotter, or card puncher.

The idea for the punched cards came from Joseph Jacquard (1752-1834), who used them to control weaving patterns in his French textile looms. Babbage used one style of card to specify arithmetical operations like adding, subtracting, multiplying, or dividing. Larger cards had data and controlled where the answers were placed in the *store*, Babbage's word for memory. He developed conditional branching, looping,

Babbage's difference engine was built in 1832 and is the earliest known automatic calculator. It was the first mechanism that included mathematical logic. It operated by turning the crank at the top.

Babbage never completed the "analytical engine," his name for a digital computer. This one was built by the Science Museum in London in 1991 to commemorate the 200th anniversary of Babbage's birth. It works just as he had designed and is accurate to 30 decimal places. The mechanical computer operates by turning the crank on the side.

and subroutines. All are important operations in modern computer programs. Although Babbage worked on it for over 30 years, he published little technical detail about his computer.

A teenager interested in Babbage's work attended an 1833 presentation he made in Turin, Italy. Ada Lovelace (1815-1852) was a mathematically proficient and dynamic young woman. The daughter of esteemed British poet Lord (George Gordon) Byron, she assumed public relations duties on Babbage's behalf. Babbage kept poor records and most people were unaware of his progress. He asked Lovelace to translate into English his Turin lectures, which had been transcribed and printed in French. She added her own insights regarding Babbage's invention, and the final document was three times the size of the original. It is only from that publication that the world found out the details of Babbage's computer. Lovelace and Babbage were close friends until Lovelace died at the young age of 35.

Babbage made many precise cardboard templates of the estimated 200,000 parts for his analytical engine. Its construction turned out to be vastly more challenging than that of his difference engine. Babbage's personality again got in the way of his invention, and the analytical engine was never built. A crude iron demonstration model was under construction at the time of his death. There is little doubt that Babbage could have constructed a legitimate computer, if only he had had the financial resources. With a bit more government funding, he could have changed the course of history. He was twice visited by the American Joseph Henry (1797-1878), the first director of the Smithsonian Institution. Henry wrote that "[Babbage] more, perhaps, than any [person] who has ever lived, narrowed the chasm separating science and practical mechanics."

Babbage and his wife had eight children. Four survived infancy. He did not remarry after his wife's death and devoted all his energies thereafter to his calculator and computer. To commemorate the 200th anniversary of his birth (1991), the Science Museum in London commissioned the construction of a Babbage-designed difference engine using 19th-century materials and machining methods. The materials used were cast iron, steel, and bronze. Modern manufacturing techniques were used, but care was taken to make parts no more precisely than Babbage could have. The difference engine cost a half million dollars to build, has 4,000 parts, weighs three tons, and is accurate to 30 decimal places.

Babbage saw too far beyond his time and his efforts were never seriously appreciated. He enjoys more widespread esteem today than he did during his lifetime. His computer work was an isolated episode in technology that did not resume development until the late 1930s. Because Babbage had such a difficult personality and worked in an unappreciated area of technology, many ordinary citizens belittled him. Museums rejected his hardware following his death. Most of his many small models of gear trains and cam arrangements were broken up for scrap. Babbage's grand-children used some as toys until they were damaged beyond repair and discarded. Today, museum curators would kill for mere fragments of those models that their predecessors refused over a century ago. Charles Babbage was the ultimate "computer pioneer."

References

Charles Babbage and His Calculating Engines by Doron Swade, Science Museum Publication, 1991.

Portraits in Silicon by Robert Slater, MIT Press, 1989.

The Making of the Micro by Christopher Evans Van Nostrand Publishers, 1981.

Joseph HENRY

It would be difficult to name a piece of machinery that does not use a transformer or electric motor. These essential pieces of modern technology were developed by a quiet, unassuming American who always maintained a dignified bearing. Joseph Henry presented useful electromagnetism to the world in 1831—but some say that the Smithsonian Institution was his greatest invention.

Smithsonian Institution Photo No. 42836

Born: December 17, 1797, in Albany, New York

Died: May 13, 1878, in Washington, D.C.

Henry came from a poor home. His father was a day laborer, and as a boy Henry went to live with his grandmother in a nearby county. He attended the local school and worked as a clerk in a local store. In his early teens, he returned to Albany to live with his recently widowed mother and became a jeweler's apprentice. Henry also started to pursue a career as a stage actor and even wrote two plays, but he changed his mind after reading a popularly written science book. Henry later said that the book was not particularly well written or profound. It was merely the first one on a technical subject that he had come across.

Henry worked his way through a local college by teaching and serving as an administrative assistant. His work and knowledge were so highly regarded that the Albany Academy offered him a full-time teaching position in 1826. There, he began his investigations into the emerging field of electricity and its relationship to magnetism. He was the first American to experiment with electricity in any important way since Benjamin Franklin had done so, almost 80 years earlier.

Using homemade batteries, Henry tried various lengths and diameters of wire wrapped around a soft iron core to produce electromagnetism. The wire available was not insulated, so Henry began by using a single loose wrapping separated from the iron core with a layer of wax. The resulting magnet could pick up only a few ounces. Henry became the first to experiment with layers of wire, which required insulation. Insulated wire was not easy to come by in those pre-electricity days. Henry tore up a silk petticoat and spent much time engaged in the boring task of wrapping insulation around wires. He found that he could wind the wires in sections, allowing him to join them in different combinations. When he connected the sections in parallel with a battery, the magnet became quite strong. Even a weak battery could produce a great deal of magnetism in this way. In 1831, Henry made a 21-pound electromagnet that could lift 750 pounds. Later that year, using the current from a storage battery, he made a one-foot-high horseshoe-shaped electromagnet that lifted one ton of iron. People at the Penfield Iron Works in Crown Point near Fort Ticonderoga heard of the powerful magnet. To help them separate iron from iron ore, they asked Henry to make two of the magnets. Shortly afterward, Crown Point was renamed Port Henry.

The next year Henry married, and he took a job at the College of New Jersey in Princeton. He conducted experiments on smaller electromagnets that could be used for fine control, and in 1835 he made the first electromagnetic relay. In effect, he had invented the telegraph nine years before Samuel Morse. He later freely gave advice to Morse, who was woefully ignorant of electrical principles. Henry never patented any of his inventions because he believed that technical discoveries should benefit everyone. Others—including Thomas Jefferson—shared his view.

In 1830, Henry had discovered and worked out the theoretical problems associated with induced electrical currents such as those connected with transformers. He worked at the Albany Academy then, and he had a heavy teaching load. Preoccupied with daily work, he put off publicizing his results. Michael Faraday discovered the same effect a few months later in England, and Faraday did publish his findings. Henry gracefully handled his disappointment over not being recognized as the discoverer. Faraday went on to create the electrical generator and was one of the foremost 19th-century British scientists. The unit of capacitance, the *farad*, is named in his honor. Like Henry, Faraday came from a poor family, had little schooling, and needed to go to work at an early age. The two met in 1835 when Henry visited England. Faraday proposed that the Royal Society of London, the world's most prestigious scientific organization, award Henry its coveted Copley Medal. Benjamin Franklin was the only other American so honored.

Henry continued investigating electricity and described the construction of an electric motor. He called the first one a "philosophical toy." It was a bar-shaped electromagnet positioned horizontally that rocked back and forth when the current reversed. From that humble beginning, Henry expanded the world's understanding of electric motors. Inexpensive electrical power was just becoming available and Henry's motor provided a method for converting electrical energy to mechanical energy. The unit of magnetic inductance, the *henry*, is named in his honor. Professionally active and respected by everyone, Henry helped organize the American Association for the Advancement of Science. He was an original member of the National Academy of Sciences and served as its president for 10 years.

Henry left Princeton for Washington, D.C., because of an 1829 bequest made by James Smithson, a British citizen. Smithson was a wealthy philanthropist who had a strong interest in science. In his will, he left $515,000 "for the purpose of founding an institution at Washington to be called the Smithsonian Institution for the increase and diffusion of knowledge." Because of legal technicalities, the organization was not founded until 1846 when the board of directors asked Henry to become the first secretary and director of the Smithsonian Institution. Always pleasant, approachable, and highly respected, he served in that position for the rest of his life, over 30 years.

The fledgling organization searched for a purpose and a direction. Many people had ideas at the time, such as having the Smithsonian sponsor a school of steam engineering or an extensive lecture series. Henry resisted all these ideas and developed a unique goal. He considered it important to distribute technical and scientific information among all people working in technology. He proposed to assist people "in making original researches, to publish these in a series of volumes, and to give a copy to every first-class library on the face of the earth." The Smithsonian's board of directors approved Henry's plan, and the National Geographic Society called it his greatest invention. Henry, more than anyone else, helped with the invention of American science.

The Smithsonian's impact on American technology was enormous. Henry interviewed young up-and-coming scientists and recommended them for jobs. He kept track of who was doing what and put investigators in contact with each other. In his later years, Henry once met with the youthful Alexander Graham Bell, who was anxious to discuss his experiments. Despite having a severe cold, Henry listened patiently and offered encouragement. Bell later said that had Henry not been so understanding, "I should never have gone on with the telephone."

No American better served the technology of his time, and Henry's death was viewed as a great loss to the public. A memorial service was held for him in the Hall of the House of Representatives on January 16, 1879. It was attended by the president and his cabinet, both houses of Congress, members of the Supreme Court, and many other distinguished people. Few Americans have been so honored by their government.

References

Those Inventive Americans, National Geographic Society, 1971.

Famous American Men of Science by J. G. Crowther, Books for Libraries Press,1969.

Joseph Henry: His Life and Work by Thomas Coulson, Princeton University Press, 1950.

Dictionary of American Biography, Charles Scribner's Sons Publishers, 1932; with supplemental updates.

National Cyclopedia of American Biography, James T. White & Co. Publishers, 1891; with supplemental updates.

McGraw-Hill Encyclopedia of Biography, McGraw-Hill, 1973.

Asimov's Biographical Encyclopedia of Science and Technology, by Isaac Asimov, Doubleday & Co. Publishers, 1964.

W. H. Fox TALBOT

Born: February 11, 1800, in Melbury, England

Died: September 17, 1877, in Lacock Abbey, England

Songwriter and singer Jim Croce (1943-1973) recorded a lovely song just before he died in an airplane crash. Titled "Time in a Bottle," it included the phrase: "If I could save time in a bottle, the first thing that I'd like to do is save every day." Although it's not possible to literally accomplish what Croce's popular song suggested, people can figuratively get close. Photography lets us stop time by recording life as we see it. The technology has been applied by the technical world in a variety of ways. Satellite photography helps road builders find the best route in undeveloped areas. Companies employ technical photographers to show customers and suppliers the details of a project's progress. The quality of a building's insulation can be evaluated by infrared photography.

Like almost all technologies, photography did not spring fully formed from the mind of a single person. It was an evolutionary process marked by milestones. In 1727, German scientist Johann Schulze discovered that silver salts turn dark when exposed to light. Joseph Niepce (1765-1833) photographed the first crude image in France in 1822. His countryman Louis Daguerre (1787-1851) developed the first practical method for producing permanent images in the 1830s. Daguerreotypes were particularly popular for family portraits. But they were a one-time image, like an oil painting is. People could not make additional copies from a daguerreotype. William Henry Fox Talbot established photography's most important milestone when he invented the method now used all over the world. Talbot was the first person to produce a positive image from a negative.

The Science Museum/Science & Society Picture Library

Talbot was born near Dorchester in southern England. His father was a British army officer who often lived beyond his means. He died when Talbot was an infant. His mother was a member of an aristocratic family, but they had fallen on hard times. They owned a large estate but were forced to rent it to a wealthy member of Parliament to pay their bills. Although not poor in the traditional sense, Talbot and his mother often moved from the home of one friend to another. His mother married admiral Charles Feilding when Talbot was four and he settled down to a more stable childhood. He grew up with two half-sisters, Caroline and Horatia. Talbot had a close lifelong relationship with them and his stepfather.

Talbott attended boarding schools, where

his grades always put him near the top of his class. His methodical approach to problem solving impressed his teachers. He entered Trinity College at Cambridge University and graduated in 1821 after studying classical languages and mathematics. His family was now financially secure, and Talbot did not have to work for a living. He was in line to inherit the ancestral home at Lacock Abbey near Chippenham, though it was still under lease. Talbot spent his time traveling while he studied his favorite subjects of mathematics, astronomy, optics, and photography. He published more than 50 scientific papers and eventually took out 12 patents.

Talbott's friends included Charles Wheatstone (1802-1875), who invented the first practical telegraph and Charles Babbage (1791-1871), who worked on an early computer. Another friend, astronomer John Hershel (1792- 1871), later suggested the word *photography*. It comes from two Greek words meaning "light" and "painting." The Royal Astronomical Society selected Talbot for membership in 1822 as did the prestigious Royal Society in 1831. Very few people were accepted into the Royal Society at the relatively young age of 32. But Talbot's insightful work with math caught the notice of the members. He had written extensively on the subject for several years.

Lacock Abbey finally became available in 1827. Talbot moved in with his mother, stepfather, and half-sisters. He remained there for the rest of his life. He married Constance Mundy in 1832 and they eventually had four children. Talbot became interested in photography during their honeymoon in Italy. He used the camera obscura technique to make drawings of Lake Como. A lens focused the scene onto a sheet of paper and Talbot sketched the image poorly with pen and ink. Realizing that he lacked artistic ability, Talbot wondered if the image could imprint itself onto a special paper. When he worked out a method, his wife became almost as interested in the subject as he was. She later became the world's first woman photographer.

As scientists and technologists had done for centuries, Talbot built on the work of others. He studied their experiments to produce his own light-sensitive materials. In a completely darkened room, he brushed ordinary paper with a solution of silver and iodine compounds. He placed the dried paper, with delicate ferns and flower petals on top of it, under glass. Exposing these items to sunlight produced dark reversed images that are now called *photograms*. No camera was involved and Talbot produced only a silhouette. He took that paper negative, laid another sheet of sensitized paper on top of it, and took it outdoors into the sunlight. The sandwiched combination took more exposure time because light had to penetrate the paper negative. He used 100 grains of silver nitrate dissolved in six ounces of water as a developer and saltwater as a fixer. Talbot's longtime assistant, Nicholaas Henneman helped him in his work.

The next step involved using the procedure with a camera to record shades of gray. Talbott ordered the construction of a light-tight box. The box had a lens in front and a sliding back for focusing. In August 1835, Talbot placed a sheet of light-sensitive paper inside the camera. He aimed the lens at a latticed window in Lacock Abbey and opened the shutter for several minutes. He processed the paper negative and then made a print from it with sunlight. It was the first negative-to-positive photograph ever made and was surprisingly sharp. On one print, Talbot wrote, "When first made, the squares of glass [in the window], about 200 in number, could be counted with the help of a [magnifying] lens."

Talbot originally called his process *photogenic drawing,* but changed its name to *calotype* from the Greek *kalos*, which means

An early camera used by Talbot, 1840. It had a simple lens with an inspection hole in the upper right to check that the image was centrally located on the sensitized paper and correctly focused.

Copy of the earliest paper photograph in existence, made in 1835. Celluloid was not used as a carrier for light-sensitive photographic emulsion until after 1870.

"beautiful." He improved the process and took out a patent in 1841. His developer was a combination of silver nitrate and gallium, which increased the sensitivity of the paper. Talbot fixed the image with sodium thiosulfate, the same chemical used today. Although calotypes were grainier than daguerreotypes, they could be used to make multiple copies and larger prints. Talbot knew that natural light might not always suffice for exposing sensitized paper and was the first to experiment with flash photography. He fastened a piece of newspaper to a spinning disk in 1851. Then he discharged a spark to expose his photographic emulsion. A portion of the paper was readable in the final print.

Talbot himself took hundreds of photographs, often with a camera of the type shown in the photograph of him at the beginning of this chapter. His favorite produced a 4-1/16-inch square negative. A book he wrote and published in 1844 titled *The Pencil of Nature* was the first book illustrated with photographs. Talbot established a photographic studio in Reading, but calotypes had only limited commercial success. After 1851, more detailed images were possible with wet solutions on glass plates. But the calotype process was instrumental in Talbot's receiving the 1842 Rumford Medal, a precursor to the Nobel Prize, from the Royal Society.

A man of diverse interests, Talbot was elected to the British parliament in 1832. He held patents for internal combustion engines and the chemical plating of metal. He had an appreciation for nature and many of his photographs involved natural settings. He also enjoyed traveling and reading about history. Talbot was fascinated when the Rosetta Stone was used to decipher Egyptian hieroglyphics in the 1820s. Possessed of a natural flair for languages, he translated written works from Assyrian and other ancient languages. He wrote five books on the subject and had 62 articles published. Talbot died at his home at the age of 77.

Talbot not only developed photographic technology, he used it brilliantly. His images had a modern character to them. They were not stiffly posed indoor portraits like daguerreotypes became. His pictures were often interesting candid shots of people in typical outdoor settings. Talbot tried to include natural elements in his pictures. He wrote poetically and philosophically about the potential of photography. Few early photographs are more compelling than those he took in the mid-19th century.

References

Fox Talbot by John Hannavy, Shire Publications, 1976.

Fox Talbot and the Invention of Photography by Gail Buckland, David R. Goodine Publisher, 1980.

The History of Photography by Helmut Gernsheim, Thames and Hudson Publishers, 1955.

Charles WHEATSTONE

It is common for an inventor to stay with one technology and become an expert in that particular field. But some develop several different technical accomplishments. Joseph Swan (1828-1914) of England invented an incandescent lamp, photographic paper, and synthetic fibers. Gustave Eiffel (1832-1923) of France designed and supervised the construction of the Eiffel Tower. He also worked on aeronautics and constructing the Panama Canal. American-born Benjamin Thompson (1753-1814) established the field of ballistics and improved heating and cooking equipment. He also built a large city park in Munich, Germany, named the English Garden.

More unusual was Austrian-American Michael Pupin (1858-1935). He invented telephone and x-ray equipment and won a 1924 Pulitzer Prize for his autobiography. Charles Wheatstone also had diverse accomplishments. Trained as a musician, Wheatstone invented the small bellows-operated concertina and the stereoscope. He devised a secret code used to rescue a future American president. Wheatstone had a working telegraph system five years before Samuel Morse (1791-1872). But he is best known for an electrical circuit that accurately measures resistances.

Wheatstone was born about 100 miles northwest of London. His father was in the retail music trade and moved the family to London when his son was four. The elder Wheatstone taught music, and he made and sold instruments. The home environment introduced the young Wheatstone to music, sound transmission, and wave propagation. He received his primary education at a private school and was apprenticed in 1816 to

The Science Museum / Science & Society Picture Library

Born: February 6, 1802, in Gloucester, England

Died: October 19, 1875, in Paris, France

an uncle in the music business. He had no formal technical instruction.

After his uncle's death in 1823, Wheatstone and his brother William took over the business. They specialized in making flutes and other instruments that used moving air. Wheatstone invented and patented the concertina in 1829. His company had others produce the parts, which they assembled and sold as completed concertinas. Wheatstone was an experimenter at heart. He built a small pipe organ to analyze air columns. He wanted to understand the properties of tone in terms of vibration. He spent much of his time trying to send sound through solids. He built a small harp he called an "enchanted lyre." He suspended the harp from a ceiling with a wire the "thickness of a goose quill." A piano on an upper floor

Wheatstone ABC telegraph of 1860. Received letters were read on the top dial. To send a letter, the operator pressed the proper button and turned the crank.

was attached to the wire and caused the harp to make music when the piano was played. Wheatstone worked at sending music and speech over considerable distances. But he concluded that electrical methods were more promising than mechanical methods. He began investigating the field of electricity but did not neglect his other varied interests.

Wheatstone measured the speed of electrons in a conductor and tried to slow down a spark. He examined light given off by burning metals, a practice we now call *spectrum analysis*. He invented a small telescope-like polar clock that used the sun and a prism to indicate the time. He invented the stereoscope in 1832. Resembling a pair of binoculars, it allowed a person to view two images at the same time in a way that made them appear three dimensional. Wheatstone occasionally wrote technical papers and made presentations. In 1834, he was appointed professor of experimental philosophy at King's College in London. At first a part-time position, it developed into a full-time one. Wheatstone married Emma West in 1847 and they had five children.

Because Wheatstone was so shy that he rarely appeared in front of a class, he never did much teaching. As a youngster, he qualified for a French award. But he forfeited the prize because he was too bashful to recite a speech in front of an audience. In his adult years, he spent most of his time with laboratory work.

Wheatstone and Cooke five-needle indicating telegraph of 1837. For each letter received, two needles deflected a small amount and pointed at a common letter. The top letter, for example is "A." An operator read that letter when both the far left and far right arrows pointed toward "A."

Stringing several miles of wire in the college corridors, Wheatstone experimented with different telegraph methods. He teamed up with William Cooke (1806-1879) to invent a practical system. Michael Faraday (1791-1867) introduced the two men. Cooke was a professor of anatomy at Durham University and had no formal technical education. His travels made him aware of the value of instantaneous, long-distance communication. Since he was having little success trying to achieve it, Faraday advised him to contact Wheatstone. The two pooled their ideas and together invented an indicating telegraph in 1837. Instead of using dots and dashes, their telegraph used five needles to point out specific letters on a diamond-shaped lattice. Cooke and Wheatstone received a joint patent in 1837. Theirs was the first practical telegraph.

The indicating telegraph used six wires, one for each needle, and a return wire. Sending a letter, such as "F," caused two needles to deflect slightly. They would point in the direction of two lines. The letter "F" lay at their intersection. Wheatstone and Cooke established the Electric Telegraph Company. Their system linked the port city of Liverpool with the manufacturing city of Manchester in 1839. This was the world's first public telegraph service. Operators required little training and could send about 22 words per minute.

Railways showed particular interest in rapid communication. George Stephenson (1781-1848) built the first major line in 1830 between Liverpool and Manchester. Early railway lines used only a single set of tracks and it was necessary to communicate about late departures or damaged equipment. British railways adopted the Wheatstone-Cooke system extensively. By 1852, more than 4,000 miles of telegraph lines were in use throughout the country. But Wheatstone and Cooke had a personal conflict and separated in 1845.

Wheatstone went on to develop another type of indicating telegraph that he called the *ABC*. Using Morse code took great skill and Wheatstone felt the future of telegraphy lay in unskilled operators. It was a reasonable assumption based on the methods used by factories. In Wheatstone's ABC system, the telegraph operator pressed a button corresponding to the desired letter. Cranking a handle sent electrical impulses. A dial rotated at the destination and indicated the transmitted letter. Wheatstone intended the system for use on private business lines, similar to a modern telephone. Like Wheatstone's joint design with Cooke, the ABC telegraph lost out to the simpler, single-wire Morse equipment.

In 1860, Wheatstone invented a printing telegraph. It was based on the Morse code, which had gained universal popularity by that time. Operators punched Morse code dots and dashes onto narrow paper tape. The tape then rapidly ran through a clock drive mechanism and sent a high-speed signal along the wire. Thin metal fingers sensed holes in the paper and opened or closed relay contacts as indicated.

Wheatstone is most commonly identified with the Wheatstone Bridge for measuring resistance, a circuit he did not develop and one for which he never claimed credit. It was actually devised in 1833 by his close friend Samuel Hunter Christie (1784-1865). Wheatstone's name was associated with the bridge circuit because he often used it in his experiments. He called it a "differential resistance measurer" and said in 1843, "To Mr. Christie must, therefore, be attributed the first idea of this useful and accurate method of measuring resistances."

A tireless worker, Wheatstone also had other notable accomplishments, including invention of the variable resistor in 1843. He improved dc generators by making one that delivered a more constant current. In 1844, he was the first to experiment with underwater cables. He invented the chronoscope, an instrument to measure small time intervals. Wheatstone discovered in 1867 that the earth's magnetic field was strong enough to start a generator. Previously, batteries were necessary. Queen Victoria knighted Wheatstone in 1868 and

Wheatstone telegraph transmitter of 1870. Paper tape was prepunched according to standard Morse code. This clockwork mechanism then automatically sent the message at a high transmission rate.

A simplified example of the Playfair cipher developed by Wheatstone in 1854:

Code word is "machine." Letters "i" and "j" are interchangeable.

M A C H I
N **E** B **D** F
G K L O P
Q **R** S **T** U
V W X Y Z

The message to be sent is "Meet Alex at museum"

1. Break message into pairs of letters: ME ET AL EX AT MU SE UM
2. Each letter pair defines a rectangle.
3. The enciphered text uses the letters at the other corners of the rectangle. For example, ET becomes DR.
4. Plain text: ME ET AL EX AT MU SE UM
   Cipher text: AN DR CK BW HR IQ RB QI
5. Send enciphered message: AN DR CK BW HR IQ RB QI

he received at least 34 awards of distinction. Wheatstone died while on a business trip to Paris in 1875.

Of all Wheatstone's accomplishments, perhaps the most colorful was his development of a secret code, or *cipher*. Cryptography was a hobby of his and he came up with a simple and reasonably secure cipher in 1854. It used a five-inch by five-inch grid of 25 letters of the alphabet. The letters "i" and "j" were assumed to be interchangeable. The cipher was named after Wheatstone's friend and look-alike Lyon Playfair (1818-1898). (The men looked so similar that even Wheatstone's wife once confused the two.) Almost a century later, Lt. John F. Kennedy's torpedo boat was destroyed in World War II combat in the South Pacific. Australian coast watcher Lt. Arthur Evans received a Playfair-enciphered radio message in August 1943. It read, "PT boat 109 lost in action in Blackett Strait two miles SW Meresu Cove. Crew of 12. Request any information." Kennedy and his crew were rescued five days later.

References

Pioneers of Electrical Communication by Rollo Appleyard, Books for Libraries Press, 1930.

The Codebreakers by David Kahn, Macmillan Publishers, 1967.

Electric Telegraphs, William Cooke and Charles Wheatstone, British Patent 7390, issued 1847.

Making of the Modern World edited by Neil Cossons, Science Museum Publications, 1992.

Ada LOVELACE

For years, computer professionals have been viewed as a colorful group. Today, they often dress more casually than others. They sometimes use words and sentence structure that seem peculiar to the average person. They have added some unusual acronyms to our language. (An *acronym* is a word formed from the first letter of a series of words.) Computer professionals used the word ENIAC in 1945 as the name of the first electronic computer. The letters stand for Electronic Numerical Integrator and Computer. The letters in the 1964 computer language BASIC stand for Beginners All-purpose Symbolic Instruction Code. The ubiquitous CD-ROM stands for Compact Disk-Read Only Memory. The American Department of Defense (DOD) introduced the country's first government-sponsored computer language in 1980. It is named Ada. But unlike ENIAC, BASIC, CD-ROM, and many others, "Ada" is not an acronym. It was the first name of a real person who played an important role in computer development.

The Science Museum / Science & Society Picture Library

Born:
December 10, 1815, in London, England

Died:
November 27, 1852, in London, England

It is sometimes hard to identify precise origins in technology and people often add qualifying words to particular accomplishments. Examples include the first *practical* incandescent lamp, or the first *patented* automobile, or the first *dependable* fountain pen. No such restriction is necessary for Ada Lovelace, one of the most picturesque individuals in computer history. She was without question the world's very first computer programmer. Lovelace wrote a program in 1843 for Charles Babbage's computer.

Although now known as Countess Ada Lovelace, her name at birth was Augusta Ada Byron. Her birthplace overlooked London's Green Park. Called Ada by her family, she was the only child of the noted British poet Lord Byron (George Gordon) (1788-1824) and his wife Annabella Milbanke Byron. Both of her parents came from financially comfortable backgrounds, though their temperaments differed considerably. Lord Byron remains Britain's most loved romantic poet and his large ancestral home of Newstead Abbey near Nottingham is open to the public. Local newspapers still print feature articles that deal with his poetry and his modern descendants. But Lord Byron led a carefree life and ignored his responsibilities. In contrast, Annabella was a calm and collected person, far more reliable than her husband. The couple's marriage lasted just

The only two packs of surviving punched cards for Babbage's never-built computer. Lovelace planned for the smaller ones to be used for mathematical operations such as multiplying. The larger cards specified numbers to be used in the calculations.

one year. The future Ada Lovelace's parents separated shortly after her birth, and she never met her father.

Byron left Britain, never to return, in part because his lifestyle was unacceptable to almost everyone in his native country. But he remembered his daughter and always displayed a picture of her. He occasionally sent her presents. In his poem "Don Juan," Byron describes a woman whose "favourite science was . . . mathematical. . . . [S]he was a walking . . . prodigy." He knew that was what Annabella expected her daughter to become. Byron wandered restlessly throughout Europe and did not remarry. He died in Greece when his daughter was eight years old.

Lovelace was a gifted youngster who displayed both artistic and intellectual abilities. Her mother was an amateur mathematician who did not want her daughter to follow in her father's footsteps. So she encouraged Lovelace to pursue interests in science and mathematics, rather than poetry. Lovelace was educated by tutors, as was the custom among the upper-middle class in the 19th century. She received special instruction from the prominent mathematician Augustus de Morgan. He was partly responsible for developing modern algebra and had a high opinion of Lovelace's abilities. She became a close friend of Morgan's wife, Sofia.

Lovelace's mother did not neglect her instruction in the arts. She learned French and entertained guests by playing the harp and violin. She particularly enjoyed ice skating and horseback riding. But she suffered from lifelong poor health and could not always actively pursue her interests. Lovelace had severe headaches at age 7 and starting at age 13 could not walk for three years. This may have resulted from measles. Lovelace spent much of that period studying mathematics and music, two surprisingly related subjects.

In the mid-1820s, Lovelace became a close friend of popular scientific writer and mathematician Mary Somerville (1780-1872). Cambridge University used Somerville's texts and Somerville College at Oxford is named in her honor. Lovelace saw Somerville as a role model and hoped to emulate her achievements. Somerville and Sophia Morgan introduced Lovelace to Charles Babbage (1791-1871) at a dinner party in 1833. They often accompanied her on visits to discuss calculations by machine. It was a topic that others found dull because they could see no use for it. Not so Lovelace. She saw it as a future mathematical concept. She also saw in it the potential for achieving the type of acclaim that Somerville had. Babbage was 24 years older than Lovelace, but the two developed a close personal and professional relationship.

Somerville introduced Lovelace to William King, whom she married in 1835. He was elevated to the title of Earl of Lovelace three years later and Ada became the Countess of Lovelace, or Ada Lovelace. In spite of domestic obligations involving their three children and her own continued poor health, Lovelace expanded her mathematical skills through self-study. Her husband encouraged her interests. She regularly corresponded with electrical scientist Michael Faraday (1791-1867), astronomer John Herschel (1792-1871), and more often with Babbage. Although she was generally a shy and private person, she was friends with author Charles Dickens (1812-1870) and telegraph inventor Charles Wheatstone (1802-1875). Babbage made technical presentations in Turin, Italy, in 1842 about the computer he was designing. An account of the lectures was published in French by L. F. Menabrea, an Italian mathematician and ambassador to France. Babbage had always encouraged women to develop scientific interests and he asked Lovelace to translate his lectures into English. Lovelace did so and, after lengthy discussion with Babbage and others, she added many of her own insights to the document.

Lovelace described the repeated use of a

set of punched cards that were similar to subroutines in modern computers. She included mathematical examples to demonstrate Babbage's machine's calculating ability. She hinted at the possibility of programming. When published in 1843, the document she produced was three times the size of the original. It showed that Lovelace completely understood the principles of computers and computer programming.

Some people feared that a machine like Babbage's might think as people do. The modern term *artificial intelligence* mirrors that thought. For those individuals, Lovelace insightfully wrote, "For this machine is not a thinking being but simply an automaton [robot] that acts according to the laws imposed upon it." She suggested that a computer might be used to compose complex music, to produce graphic images, and for scientific calculations. Those are all modern applications. Babbage gave her great praise when he wrote that she had "entered fully into almost all the very difficult and abstract questions connected with the subject." It is only from Lovelace's publication that the modern world knows the details of Babbage's computer.

Babbage then asked Lovelace to use her skills to write a computer program. She worked out a procedure to calculate Bernoulli numbers, a complicated chore. The numbers are for specialized calculus operations. The activity of writing the program was only a mathematical exercise. Since Babbage's computer was never built, it was not possible to test Lovelace's program at the time. But it would have worked. When used with modern computers, her program gives the expected values.

Commemorative plaque inside St. Mary Magdalene Church in Hucknall, England, where Lovelace is interred next to her father, Lord Byron.

Lovelace and Babbage wanted to bring the computer to the world. But it would have required a huge amount of money to make its 200,000 close-tolerance parts and carefully assemble them. The British government showed no desire for financial involvement in such a project. Unfortunately, Lovelace and Babbage then decided to use their mathematical skills in a risky manner. They developed a gambling system that they thought could provide the required funds. But their horse-racing system did not work and both lost large sums of money. Lovelace died heavily in debt in 1852, at only 36 years of age.

Lovelace's father, Lord Byron, had also been only 36 when he died. At her request, she is buried next to him in the Byron vault inside the St. Mary Magdalene church in Hucknall. Lovelace was the last Byron to be so honored.

Air Force Colonel William Whitaker served as chair of the 1970s committee that investigated a new computer language for the DOD. Its temporary name during development was DOD-1. Before approving the permanent name Ada, Whitaker asked permission from one of Lovelace's descendants. The Earl of Lytton readily agreed and noted that Ada was "right in the middle of radar."

Lovelace's work was generations ahead of her time and few others of her era understood it. It has long been reasonably speculated that had the Lovelace and Babbage horse-racing system made money, the world might have had computers in the mid-19th century.

References

Ada—Enchantress of Numbers, narrated and edited by Betty Alexandra Toole, Strawberry Press, 1992.

The Computer from Pascal to von Neumann by Herman H. Goldstine, Princeton University Press, 1972.

Lord Byron's Wife by Malcolm Elwin, Harcourt, Brace & World, 1962.

Christopher SHOLES

Born: February 14, 1819, in Morresburg, Pennsylvania

Died: February 17, 1890, in Milwaukee, Wisconsin

By the middle 1800s, several inventors had patented a number of different styles of typewriters. They called them "writing machines," and most were cumbersome and hard to use. One design required securely clamping a sheet of paper to a large roller. The user then positioned letters by hand on a circular metal frame over the paper and made impressions with a plunger. The typewriter was necessarily complex, and all early designs proved so complicated and tedious to use that potential purchasers showed no interest in them. In 1868, however, one design came to dominate typewriter construction with two unique ideas: The linkage mechanism resembled familiar piano keys, and the paper carriage moved one space after each letter was printed. The person who developed this first modern typewriter was Christopher Latham Sholes, a printer, newspaper editor, postmaster, and politician.

Sholes was born on a farm. At the age of 14, he left school to apprentice with a local newspaper, the *Danville Intelligencer*. His interest in inventing began during his apprenticeship. The young Sholes devised a method of addressing newspapers by printing subscriber names in the margin. After completing his four-year apprenticeship, Sholes moved with his family to Wisconsin. He spent the rest of his life there.

Smithsonian Institution Photo No. 38142

Within a year of moving to Wisconsin, Sholes became the editor of the territorial legislature's journal, the *Wisconsin Enquirer.* During his early working life, Sholes worked for five newspapers, usually as an editor. He also served for a while as postmaster of Kenosha. After Wisconsin became the thirtieth state, Sholes became active in politics, in addition to his work as postmaster and journalist. He was poorly suited to the role of politician, being soft spoken and suffering frail health. Still, he was elected to and honorably served two terms as senator and one in the state assembly.

Sholes's extremely busy life kept him from devoting much time to inventing until 1862. That year, President Abraham Lincoln appointed him to serve as Milwaukee's Harbor Customs Collector. The job was not time consuming, and it allowed Sholes to devote more energy to refining the typewriter. An article on writing machines by inventor Ely Beach in *Scientific American* encouraged him. After discussing the article, Sholes and two friends decided to try to invent an easier-to-use, more efficient typewriter.

Sholes worked on the typewriter with a machinist named Samuel Soule. Sholes and Soule had already patented a device for automatically numbering book pages. They created their invention in a small machine shop they rented on the second floor of a brass foundry. They shared space with a third inventor, Carlos Glidden, a lawyer and court reporter. With Sholes as their unofficial leader, the three inventors collaborated on a typewriter and received a joint patent in 1868. Their invention is regarded as the first modern typewriter. They also devised the four-row universal keyboard in use today. It is often called the QWERTY layout, after the first six letters in the upper row of letter keys. The patent model had only 11 keys and was not intended as a production unit.

Sholes and his partners had an excellent product, but they soon found there was no market for it. They offered their typewriter to bankers and authors but made few sales. Bankers disliked it because they felt that typewritten contracts could be too easily altered, rendering them illegal. Although Mark Twain purchased a typewriter, there were in general too few authors to provide a reasonable market. No one considered the massive business possibilities of the typewriter. The three inventors continued to work on improvements, especially on the keyboard arrangement. One early experiment involved a single row of 44 characters, with the letters in alphabetical order.

In the Sholes typewriter, each letter was at the end of a type bar that swung in an arc on its way to the paper. Sometimes the bars collided with each other and jammed—a problem particularly for fast typists. Sholes tried many arrangements. He decided to locate the most frequently used letter combinations as far apart as possible. This would tend to slow down a fast typist and minimize jamming the type bars. The result was the QWERTY, or universal, keyboard. Mark Twain described the QWERTY keyboard arrangement as a "curiosity-breeding little joker."

Improvements to the partners' 1871 model made it the first to look like the typewriters we know. This model had a four-row keyboard and a horizontal rubber roller to hold the paper. The typewriter was mounted on a table, and a lead weight hanging from the end pulled the roller between keystrokes. A push on a foot pedal returned the paper and moved it down one line. Sholes became the machine's champion and tried without success to sell the improved version. He gave many personal demonstrations to bankers and stock brokers. But the mild-mannered man was not well suited for the job, and he made only a few sales. An 1872 front-page article in *Scientific American* generated some customers but not enough to make a difference.

Sholes's typewriter patent model is on public display at the National Museum of American History in Washington, D.C.

With little money coming from the invention, Soule and Glidden decided to sell their rights to Sholes. Sholes patented additional improvements, mortgaging his home to finance the work. He was the only one who had faith in his design, and he persisted in working on it as long as he could. Finally, in 1873, poor sales forced him to sell out to Philo Remington for $12,000. The Remington Arms Co. wanted a new product to manufacture after making guns for the Civil War. It had excellent manufacturing machinery and skilled machinists, who perfected Sholes's invention. Using a well-established sales force, Remington introduced the successful and expensive $125 Remington-Sholes typewriter in 1874.

After purchasing the patent rights, Remington hired Sholes and two of his sons as consultants. Advanced age and delicate health cut into Sholes's productivity with inventions. He took out his last patent application in 1889. It was for an even more improved keyboard. Sholes lived to the age of 71, but he was quite weak during his last nine years. He did much of his work propped up in bed. He devoted much of his time to the perfection of the typewriter and lived to see its use established throughout the world.

The QWERTY keyboard arrangement Sholes developed is not the most efficient, though it has remained standard since the 19th century. To date, the most serious attempt to improve it was launched by August Dvorak in the 1930s. With a grant from the Carnegie Foundation, Dvorak and his brother-in-law William Dealey conducted studies that showed that typing students learned more rapidly with the Dvorak simplified keyboard. They also noted a reduction in errors. Dvorak and Dealey received a patent for the keyboard in 1936. Like the QWERTY arrangement, it has four rows of keys. The characters "?,.PYFGCRL/" are in the first row of letters. The Dvorak arrangement has not gained much popularity and it appears on few keyboards today.

References

The Early Word Processors by Carrol H. Blanchard, Educators Project IV, Lake George, NY, 1981.

American Science and Invention by Mitchell Wilson, Bonanza Books, 1960.

Typewriting Behavior by August Dvorak and others, American Book Co., 1936.

Dvorak typewriter layout display, National Museum of American History, Washington, DC.

Dictionary of American Biography, Charles Scribner's Sons Publishers, 1932; with supplemental updates

National Cyclopedia of American Biography, James T. White & Co. Publishers, 1891; with supplemental updates.

Cyrus FIELD

Born:
November 30, 1819, in Stockbridge, Massachusetts

Died:
July 11, 1892, in New York, New York

To obtain European news during the mid 1800s, U.S. journalists frequently interviewed passengers leaving ships. Communication was so haphazard that freighters cruised from port to port searching for cargo. The situation greatly improved in 1866 when Cyrus Field established the first permanent transatlantic telegraph cable between Europe and North America. The project wasn't trouble free, and Field endured more than a decade of expensive cable-laying failures. Each of his 40 transatlantic crossings made him seasick.

Field was the youngest of seven sons. At the age of 15, unlike his older brothers, he decided not to attend college. (One of his brothers later became a chief justice on the U.S. Supreme Court.) Leaving home with only $8, Field found employment as an errand boy in New York City. Three years later, he went to work at a wholesale paper dealership. Field became a junior partner at 21 and almost single-handedly worked the organization out of bankruptcy. He amassed a personal wealth of $250,000. Yet long hours and six-day work weeks had left him exhausted and in frail health. He retired from the paper business at 33.

Then, Field met British engineer Frederick Gisborne, who was working on a telegraph line across the wild terrain of Canada's Newfoundland. By obtaining information from ships that first stopped at St. John's, Newfoundland, his company got news several days before the ships would arrive with it in Boston and New York. Discussions with Gisborne gave Field the idea of laying a connecting cable across the Atlantic Ocean. He also discussed possible telegraphic transmission problems with Samuel Morse. Morse

Chicago Historical Society

Cyrus Field with a section of his transatlantic telegraph cable

assured Field that a signal could be transmitted over the 2,300-mile underwater distance. Field consulted a globe to verify that the shortest distance for the first transatlantic cable would require that it come ashore in Newfoundland. This would be an enormous undertaking, one on a scale unlike anything previously attempted. In modern times, a governmental body—rather than a single individual—would spearhead a comparable project.

At this point no U.S. company made underwater cable, so Field contacted a British manufacturer to design one. It started as a single conductor with seven strands of copper wire. The wire was insulated with three layers of a new type of rubber called *gutta-percha*. The insulation was rigid and strong, and for the next 70 years, nothing could match it for use in underwater cables. It came from the jungles of Malaya, and the name evolved from the Malaysian words for gum (*getah*) and tree (*percha*).

The insulated conductor in Field's cable was strengthened with spirally wrapped iron wires. In all, Field used 367,000 miles of iron wire. Hemp and tar sealed the entire cable. All the necessary 2,600 miles of cable was made ready in only six months. The photograph on the previous page shows the slender Field holding a piece of the cable alongside a globe, to symbolize the cable's ability to bind the world together.

The cable weighed 2,000 pounds per mile, and huge ships were needed to carry it. The U.S. and British governments offered support, each providing a ship with special cable-handling equipment and trained personnel. The U.S. Navy used its largest ship, the *Niagara*, and the Royal Navy used the *Agamemnon*, one of the last wooden-hulled ships in the British fleet. During the summer of 1857, each was loaded in Ireland with about 1,300 miles of cable. The plan was for the *Niagara* to begin laying cable and splice it to that on the *Agamemnon* halfway to Newfoundland. Complex machinery let out the garden-hose-sized cable as the ships sailed west at a slow 4 mph. Tension was carefully adjusted using sensitive brakes to support the enormous weight of the cable hanging thousands of feet to the ocean floor.

This sensitive galvanometer was used in Newfoundland in 1858. It received a message from Queen Victoria to President James Buchanan.

After only 335 miles, the brakes grabbed too firmly and the cable broke. It fell into the ocean, and could not be retrieved. Although the corporation lost $500,000, Field used his persuasive manner and impressive organizing skills to arrange four more attempts. The fourth succeeded temporarily. On August 16, 1858, Queen Victoria sent the first official transatlantic telegraph message to President James Buchanan. It consisted of a 90-word greeting.

Receivers at each end of the cable used unique signal amplifiers. Silk thread suspended a magnet inside a coil of fine copper wire. The thread had a mirror attached near its midpoint. Weak electrical telegraph signals arrived at the coil, causing the magnet to rotate slightly. The rotation made the mirror move slowly one way or the other. Light reflected by the mirror would move to the left or right indicating Morse code dots and dashes. This sensitive galvanometer sometimes gave incorrect readings and required confirmation from the sender. The photograph below shows the setup used in Newfoundland. At a maximum transmission rate of four words per minute, transatlantic telegraphy was much slower than land telegraphy. Queen Victoria's message took more than 16 hours to send and verify.

About 400 messages were transmitted before the cable broke on September 1. An economic depression in 1860 and the Civil War that followed delayed further efforts. Yet Field had tremendous personal drive and never lost sight of his goal. The American and British governments realized that the cable was more necessary than ever and supported Field in every way possible. Before the Civil War was over, Field had engaged the services of a huge British steamship.

At 692 feet in length, the *Great Eastern* had about five times more capacity than the next-largest ship. No ship larger was built for almost 50 years. The ship had been designed to transport 4,000 passengers to the East Indies without having to refuel. Instead, the owners decided to operate it on the shorter, more competitive Atlantic Ocean routes. The *Great Eastern* never made a profit as a passenger vessel, but its great size proved ideal for carrying all the underwater cable necessary to span the ocean.

The crew of the *Great Eastern* started laying cable from Ireland in 1865. About 600 miles from completion, the cable broke and fell two and a half miles to the ocean floor. Weak grappling equipment could not lift the broken end of the new heavier cable. Yet, even this sixth failure did not deter Field. The next year he made another attempt and that one succeeded completely. The cable was permanently completed on July 27 between Valentia, Ireland, and Hearts Content, Newfoundland. The project was called the greatest engineering feat of the century up to that time.

This event was almost immediately followed by completion of a second cable. Field ordered the *Great Eastern* back out to sea the day after it landed. Using better grappling equipment, trained personnel recovered the broken end of the cable that had fallen in 1865 and spliced it to a new section. Within days, people were literally standing in line to send transatlantic messages at $5 to $10 per word.

Field received innumerable awards and other forms of recognition. He donated most of his monetary awards to the New York Metropolitan Museum of Art. Field went on to assist with cable laying in many other countries. He also helped with the elevated railroads in New York City and devoted much time to their completion. While in his seventies, Field discovered that most of his wealth was gone through mishandling by financial advisors whom he considered friends. A bright spot came in 1890 when he and his wife, Mary, celebrated their golden anniversary. They had seven children and numerous grandchildren.

The British steamship, *Great Eastern*

References

History of the Atlantic Telegraph by Henry Field (a relative of Cyrus's), Books for Libraries Press, 1866.

Voice Across the Sea by Arthur C. Clarke, William Luscombe Publisher, 1958.

"The Cable Under the Sea" by James R. Chiles, in *American Heritage of Invention and Technology*, Fall 1987.

American Science and Invention by Mitchell Wilson, Bonanza Books, 1960.

Dictionary of American Biography, Charles Scribner's Sons Publishers, 1932; with supplemental updates.

National Cyclopedia of American Biography, James T. White and Co. Publishers, 1891; with supplemental updates.

McGraw-Hill Encyclopedia of Biography, McGraw-Hill Publishers, 1973.

Asimov's Biographical Encyclopedia of Science and Technology, by Isaac Asimov, Doubleday & Co. Publishers, 1964.

William THOMSON
LORD KELVIN

Born: June 26, 1824, in Belfast, Ireland

Died: December 17, 1907, in Largs, Scotland

The history of human achievement in science and technology can often be read in measurement units. One common temperature scale was named for Gabriel Fahrenheit (1686-1736). He was born in Gdansk, Poland, and made the first successful mercury thermometer in 1714. Another everyday scale was named for Anders Celsius (1701-1744), a Swedish astronomer. In 1742, Celsius was the first to describe a temperature scale based on water's boiling and freezing points. He recommended dividing the scale into 100 increments.

Both the Fahrenheit and Celsius scales were based on the characteristics of water. Technologists often find absolute temperature scales more useful. Absolute zero is the lowest temperature that can be reached. It occurs at -460° Fahrenheit. A scale that begins at that temperature is the Rankine scale, named for William Rankine (1820-1872) of Scotland. He was one of a handful of people who created engineering as an academic discipline. In the 19th century, engineering and technology were identical career paths.

A person who spent almost all his life in Scotland also has a temperature scale named for him. He is Lord Kelvin. Absolute zero on the Kelvin scale occurs at -273° Celsius. Lord Kelvin, born William Thomson, worked with heat and energy, or *thermodynamics*. He also invented a special galvanometer that made the 1865 transatlantic telegraph cable practical.

Kelvin was the fourth of six children in his family. He was born in Ireland, where his father was a professor of mathematics. His father had some ideas about edu-

Lord Kelvin—An Account of His Scientific Life and Work by Andrew Gray, J. M. Dent Publishers, 1908

cation and personally taught Kelvin and his brother James at home. Kelvin's mother died in 1830 and his father assumed personal responsibility for raising all the children. When he accepted a position as head of mathematics at the University of Glasgow, he moved them all to Glasgow. Kelvin's father deserves some credit for providing the environment that placed his son on the path to technical discovery and international recognition. Kelvin was a child prodigy who entered the university at the age of 10, along with his older brother who was 12. Kelvin was usually at the top of his classes and James was often second. James eventually became a professor of engineering at the University of Glasgow.

Kelvin published the first of his 661 technical papers when he was only 17. It dealt with advanced mathematics. He continued his studies at Cambridge University and Paris University in France. Kelvin specialized in mathematics, optics, and astronomy. He received an advanced degree from Cambridge in 1845. A professorship in physics became available at Glasgow and the 22-year-old Kelvin accepted the position. His father's influence was obvious. But the professorship was a good fit for Kelvin. He stayed at the University of Glasgow for the next 53 years.

This 1878 harmonic analyzer interpreted graphical records of atmospheric temperature and pressure. Part of the mechanism was invented by Kelvin's brother James.

Scottish technical education was the best in the world at the time and Kelvin worked to keep up the standards. One of his early projects involved the construction of Britain's first physics laboratory. His department offered the most advanced physics instruction available in Britain through the 1840s. Early on, Kelvin had an interest in determining the age of the earth. In an era before the discovery of measurement methods based on radioactivity, this was a difficult determination. Kelvin based his analysis on estimated cooling rates when compared with the sun. That work pointed him toward studies with heat and energy, two subjects that fascinated him for the rest of his life.

Famous Men of Science by Sarah Knowles Bolton, Thomas Y. Crowell Co. Publishers, 1889

This photo shows Lord Kelvin later in life.

James Joule (1818-1889) lived in England at the time and published papers on heat and energy that were all but ignored. Kelvin met Joule and was impressed with his work. He saw to it that Joule's technical efforts received proper attention. The two men became close professional allies, working together between 1847 and 1852.

Joule and Kelvin noticed that as gas cooled, its volume decreased by 1/273 for every degree Celsius below zero. Others had also observed that characteristic. But Kelvin proved that the volume would not drop to zero at -273° Celsius. Instead, all molecular motion stopped and the temperature could not go any lower. Kelvin called this the "absolute zero of temperature" and the name *absolute zero* stuck. He developed a temperature scale with its zero point at absolute zero. That made the freezing point of water 273° Absolute and its boiling point 373° Absolute. The absolute temperature scale was later renamed the Kelvin scale.

The results of Kelvin's and Joule's work became the basis for liquefying gases, such as liquid hydrogen. Liquid hydrogen fueled the Saturn 5 rocket that took people to the moon in the 1970s. The fuel is also used by the Space Shuttle's three main central engines.

During the late 1800s, much of Britain's technical community was working on the theory and practical aspects of a transatlantic telegraph cable. The American

Perhaps Kelvin's most significant invention, this mirror galvanometer of the 1860s saved the undersea transatlantic telegraph cable project. This one was used at the North American end of the Newfoundland-to-Ireland cable.

Cyrus Field (1819-1892) laid the first cable in 1858, but it operated only briefly. He established a permanent cable in 1865, followed by others over the next few years. Kelvin sailed on I. K. Brunel's (1806-1859) huge ship *Great Eastern* during its first successful cable-laying crossing in 1865.

There were problems associated with receiving weak electrical telegraph messages from an underwater cable that was 2,300 miles long. The person in charge of signal transmissions was E. O. W. Whitehouse. He thought the underwater cable should operate at high voltage to ensure that the electrical current was pushed along the entire distance. Kelvin quietly thought otherwise. During a test, Whitehouse's high voltages damaged the cable, but did not put it out of service. After some years of legal and technical arguments, Kelvin installed a low-voltage readout system that used his highly sensitive mirror galvanometer. About the size and shape of an ordinary flashlight battery, Kelvin's galvanometer had a tiny magnet and mirror hanging by a silk thread inside a coil of fine copper wire. Weak transatlantic signals arriving at the coil caused the mirror to rotate slightly. A beam of light reflected by the mirror multiplied the effect. Motion of the light beam to the left or right indicated Morse Code dots or dashes. Kelvin's mirror galvanometer had rescued a huge financial investment. Queen Victoria knighted him for his effort in 1866. He patented the invention and held a total of 70 patents.

Kelvin was the first scientist to grow wealthy through his many discoveries. His estate totaled £160,000 in the early 1900s. He married Margaret Crum in 1852. She became ill soon after the wedding and was an invalid until she died in 1870. Kelvin married Frances Blandy in 1874. They built a large house in Largs, near Glasgow, to complement one he had in London. Kelvin had no children from either marriage. He and Frances traveled to America in 1884, where he delivered a series of lectures in Baltimore and other places.

Kelvin once fell on the ice while playing curling, a Scottish game similar to shuffleboard. His leg was badly broken and he walked with a limp for the rest of his life. He enjoyed sailing and bought a yacht in 1870 that he named *Lalla Rookh*. He was an accomplished navigator and spent much of his spare time on the yacht. Kelvin's lifelong interest in sailing encouraged him to develop new instruments for that venture. His nautical inventions include new types of compasses, depth gauges, and a tide predictor.

After many years of productive scientific work, Queen Victoria conferred on Kelvin the title of Lord Kelvin in 1892. He was the first person to be so honored for technical achievement. He chose the

Kelvin's 1876 tide predictor was the first machine to predict high and low tides. It combined 10 tidal components, with one pulley for each. During operation, the smallest wheel spun at 1,600 rpm.

name from the small River Kelvin that flowed through Glasgow. Kelvin received almost every honor a person could. He was president of the prestigious Royal Society from 1890 to 1894. He received awards from 250 organizations. He retired in 1899 and immediately signed up for university classes as a special student. He died eight years later at the age of 83.

Kelvin did much work with theoretical analysis. His inventions included navigation and communication equipment, electrical circuits like the Kelvin Bridge and the Kelvin-Varley voltage divider, and a huge mathematical analyzer that was larger than a metal lathe. An extraordinary and likable scientific genius, Kelvin once called himself a failure because he could not "fit physical science into the engineer's concept of nature." Britain accorded him its highest final honor. Kelvin is buried next to Isaac Newton in London's Westminster Abbey.

Author Dennis Karwatka during a visit to the University of Glasgow, where Kelvin worked for 53 years.

References

Famous Men of Science by Sarah Knowles Bolton, Thomas Y. Crowell Company Publishers, 1889.

The Making of the Modern World edited by Neil Cossons, Science Museum Publications, 1992.

Joseph SWAN

Born:
October 31, 1828, in Sunderland, England

Died:
May 27, 1914, in Warlingham, England

It is not uncommon for the popular press to introduce a historical technologist as the "inventor of the bicycle," for example. They might also refer to the "inventor of the fountain pen" or the "inventor of the milling machine." What they really mean is *an* inventor of a bicycle, or *an* inventor of a fountain pen, or *an* inventor of a milling machine. Few inventions have emerged fully formed from the mind of one person. Practically all evolved through the efforts of many people. One of the best examples is Thomas Edison (1847-1931), who is often characterized as the "inventor of the electric incandescent lamp." However, Edison did not invent the lamp and never claimed that he had. He grew tired of explaining that he only made it practical.

Edison tested his first successful prototype lamp in October 1879. Early records are unclear, but it appears that Scottish schoolmaster James Bowman Lindsay may have produced light from electricity in 1835. The first patent for an incandescent lamp may have been one taken out in 1845 by J. W. Starr (1822-1847) in Cincinnati, Ohio. The lamps proved unsuccessful because of the difficulty in removing all the air from the glass envelope. One electrical pioneer might have supplanted Edison in people's minds if only he had patented his lamp in a timely manner. Joseph Swan operated several incandescent lamps using filaments of carbonized paper in 1860. That was almost 20 years before Edison succeeded with a similar filament.

Swan was born in an industrial seaport community in the northeast of England. His father sold metal fittings for ships, but he did not profit from the business. He wasted much of the family income on impractical inventions and poor business practices. The

The Science Museum/Science & Society Picture Library

younger Swan attended the Hendon Lodge School until money ran out when he was 14. But he said the out-of-school education he received was invaluable in establishing his life's work. Sunderland was a bustling city of iron works, ship building, coal mining, and supporting industries. The inquisitive teenager liked to visit the shops with his brother John and to talk with the workers.

Although Swan had hoped for another career, circumstances led him to an apprenticeship with a druggist in nearby Newcastle. But the druggist was pleasant and provided an environment for Swan to work with the new technology of electricity. He experimented with coils, condensers, and batteries. At 17, Swan heard a lecture on electricity by W. E. Staite in Sunderland. From then on, he applied much of his free time to making a practical incandescent lamp. Doing so was not a new idea and had been discussed in technical circles for more than 30 years. While growing up, Swan's long winter evenings were spent in rooms poorly illuminated by candles and oil lamps. The dim light hampered his studies after sundown and he hoped to find a better lighting method. By 1848, he had developed an incandescent lamp using a carbon filament that glowed in a vacuum in a glass envelope. But the vacuum was not good enough nor were the batteries powerful enough. Swan turned to other areas of technology.

Swan left his apprenticeship to enter the pharmacy and chemistry business with John Mawson, who would soon become his brother-in-law. Mawson was a respected businessman and introduced Swan to many community leaders. Swan was a pleasant person, who soon emerged as head of the local literary society. The charming young man met people who provided financial backing for his early experiments. His first profitable venture was in the field of dry photographic plates.

Swan saw his first photograph in 1850 and became fascinated by the process. Early photographs were exposed on glass plates that carried a wet, sticky emulsion. Swan was the first person to produce a truly dry plate. He found that if a silver bromide solution was carefully heated, its sensitivity increased. Mawson helped finance the construction of a factory in 1856. Swan's work with photography led him to investigate light-sensitive photographic paper. He patented a bromide process in 1878 and paper like his is still used to print negatives.

The shorter building in the middle is the site of Swan's early work with incandescent lamps. He was a partner in a chemistry business with John Mawson at 13 Mosley Street in Newcastle. A circular plaque at the left of the entrance door commemorates the building's technical heritage.

While manufacturing his photographic printing materials and specialty chemicals, Swan continued his lamp experiments. He paid particular attention to filament quality and high vacuum. His 1860 lamp used a quarter-inch-wide filament heated by a 50-cell battery. But the arrangement proved impractical. Swan became a well-established inventor and ultimately earned more than 70 patents. He won several awards at the Paris Exposition of 1867. Then, that same year, several tragedies befell him.

His close friend Mawson died in an accident while trying to destroy some nitroglycerine explosive. Swan's wife of five years, Frances, and their twin boys died of a mysterious fever. To escape from a house marked by sad memories, he moved a few miles away to Gateshead and stayed in virtual seclusion for more than two years. In 1871, he married Hannah White, returned to his factory in Newcastle, and was soon working again at his normal pace. Swan had a total of nine children.

The development of efficient dynamos (dc generators) and improved vacuum pumps encouraged Swan to make another serious try at the electric lamp. He did well enough financially to hire a full-time assistant, Charles Stern, to help produce light from a glowing carbon filament. Swan's lamps operated at low voltage and relatively high cur-

Swan presented this lamp to London's Science Museum in 1908. The history of this particular lamp is unclear. It may have been made in 1878 and used as an exhibit in an important demonstration in 1888.

rent. A typical 20 watt Swan lamp driven by a 5 volt dynamo required a current flow of 4 amperes. His lamp filaments had to be fairly thick to carry that much current. In addition, it would have been impractical to distribute low voltage dc and Swan seems never to have thought of doing so.

Swan made a public demonstration of a successful incandescent lamp to the Newcastle Chemical Society in December 1878. But his work with the electric lamp was primarily a hobby and Swan was slow to patent his invention. Less than a year later, Edison operated his lamp and immediately applied for a patent. He received it in January 1880.

Swan had seen his mistake and applied for a related American patent, which was granted in October 1880. He opened the Swan Electric Light Company in Britain in 1881. It proved quite successful. His lamps were the first used in the House of Commons of the British Parliament and in the British Museum. Swan once received an order from America for 25,000 lamps. Edison brought a patent infringement suit against Swan and they settled out of court in 1883, forming the Edison and Swan United Electric Company. The company dominated Great Britain's electric light manufacturing business for many years.

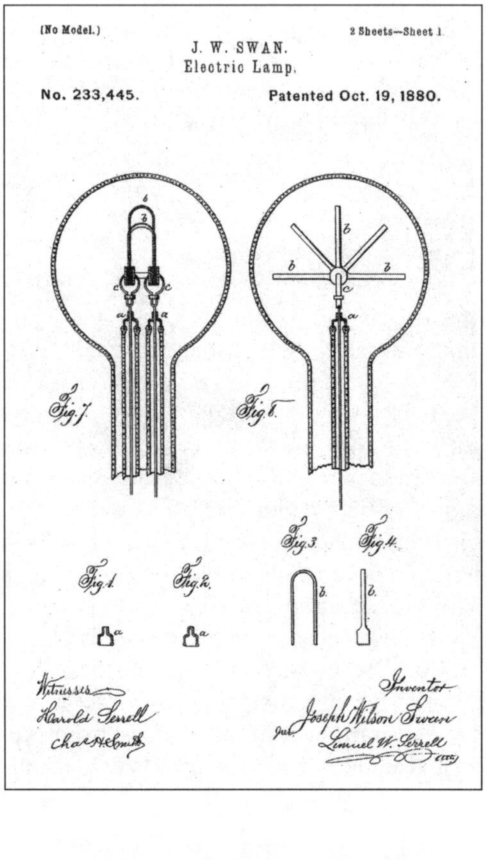

This drawing is from an American patent Swan took out in 1880. Although still using a thick high-current filament, the glass envelope and connections clearly show the influence of Edison's globe design. Swan's earlier lamps had resembled short bananas with electrical connections at each end.

Swan's fertile mind also produced the first synthetic thread and fabric. While looking for a suitable filament, he produced a thick nitrocellulose solution. It was a dangerous mixture, chemically related to nitroglycerine. Swan developed a method of forcing the hot, viscous solution through a grid of small holes into a cooling liquid. He carbonized the resulting threads and used them as lamp filaments. In a related activity, Swan treated the threads to eliminate their flammability, wove them into fabric, and made some small textile articles. He called the material *artificial silk* but did not exploit the idea. The fabric was related to rayon, which was first commercially produced in about 1910.

Swan was a dignified and modest man who lived a long and productive life. He received many honors and served as president of three professional organizations. He moved to London to be near his company and then to the south of England during his declining years. He was a poetic writer of lively letters and well liked by everyone. He died of a heart ailment at the age of 85.

In spite of the closeness of the incandescent lamp race, most historians credit Edison with inventing the first practical one. His lamps were more efficient and used hair-thin high-resistance filaments instead of the thicker low-resistance strips used by Swan. Edison also supported his invention with an electrical generating and distribution network. His 1882 Pearl Street Station in New York City was the world's first light plant. It produced 110 volts of dc electricity for 59 original customers.

Edison regularly researched the technologies with which he worked. He almost certainly knew about Swan's earlier experiments. And Edison was quick to settle his 1883 patent infringement suit. Without Swan's pioneering work, Edison might not have succeeded with his own incandescent lamp.

References

Sir Joseph Swan and the Invention of the Incandescent Electric Lamp by Kenneth R. Swan, Logmans, Greene, and Company Publishers, 1946.

Inventions That Changed the World, Reader's Digest Association, Ltd., 1982.

George WESTINGHOUSE

Born:
October 6, 1846, in Central Bridge, New York

Died:
March 12, 1914, in New York, New York

Only a few teenagers have ever received a patent. Even fewer people in their twenties establish a corporation based on a single invention. George Westinghouse did both. He patented a commercially unsuccessful rotary steam engine at the age of 19. But his railroad air brake was another matter. During a dramatic demonstration in 1869, his invention impressed many influential people. They helped the 23-year-old Westinghouse establish the Westinghouse Air Brake Co. with an initial capitalization of a half million dollars.

Westinghouse was one of nine children born into a large family in eastern New York. His father, a manufacturer of agricultural implements, introduced all his children to the technical aspects of production. At the age of 10, the younger Westinghouse worked in the machine shop each day after school. A generally disagreeable youngster, Westinghouse never showed interest in learning the use of tools. He was prone to temper tantrums and continually disrupted his school classroom. He disliked his studies and at the age of 16, asked his parents for permission to join the U.S. Navy. The earlier enlistment of two of his older brothers and Westinghouse's difficulties at school encouraged his parents to consent. Westinghouse subsequently served on several ships and attained the rank of Acting Third Assistant Engineer.

Westinghouse returned home in 1865. Family members immediately noticed the maturing effect that military service had had on him. He had learned self-discipline and developed a sense of responsibility. Those qualities served him well when he returned to work in his father's machine shop, taking on the rotary steam engine as his first project. Westinghouse got the idea for the engine while serving on steam-powered ships. He received patent number 50,759 for the engine three weeks after his 19th birthday, but the engine proved impractical. This early experience did, however, lead him to work on steam turbines later in his life.

Westinghouse Electric Corp.

Westinghouse read that Europeans used compressed air to operate drills and hammers. He was particularly fascinated by the use of compressed air in constructing the Mount Cenis Tunnel through the Italian Alps. After witnessing the collision of two trains, Westinghouse wondered if he might also use compressed air to stop trains.

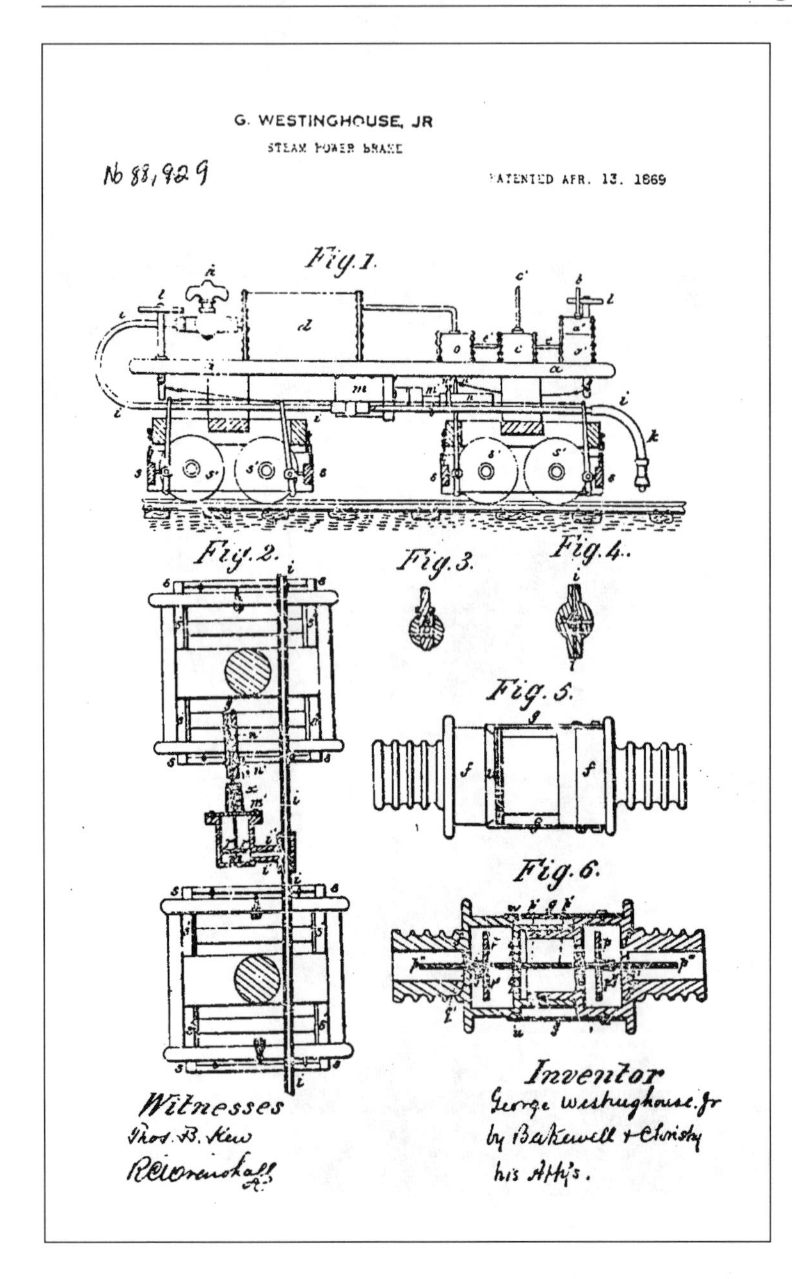

Drawing from Westinghouse's original air brake patent

Before 1869, trains were usually no more than five cars long and they traveled no faster than 30 mph. The method used to stop them made the speed limitation necessary. Each car had its own brake operator. On hearing a whistle signal from the locomotive engineer, the brake operators on each car turned brake handles. The engineer did not directly control the brakes, and it usually took about a mile to stop a train. If there was an emergency on the tracks ahead, it was often impossible to stop in time to avoid a problem. Because of this difficulty, and the existence of poor signaling systems, trains sometimes collided. They also occasionally ran off the tracks and crashed because of washed out bridges or defective rails. The engineer might see these hazards far ahead but still not be able to stop in time. Because of poor brake design, railroad travel was not as safe as it might have been. By 1867, there were more than 600 patents addressing the concern. Many inventors tried using steam pressure, but the steam condensed to water in brake lines and proved ineffective.

Westinghouse was the first to use air pressure. He invented and patented a brake that operated from compressed air at 60 psi. He designed a compressor installed near the engine that used steam power to develop air pressure. But Westinghouse found himself unable to interest others in this new idea. Financier Cornelius Vanderbilt turned the young man down for trying to "stop a train with wind." Nonetheless, Westinghouse arranged a test with the Panhandle Railroad in Pennsylvania. He installed air brakes on four passenger cars and brake controls in the locomotive, borrowing all the money he could to personally pay for the work. He scheduled a test to take place in April 1869 over 35 miles of open country between Pittsburgh and Steubenville, Ohio. Since brake operators would not ride on the train, he took special precautions to keep people from crossing the tracks.

Before the start of the test, the locomotive pulled out of Pittsburgh's Union Station with Westinghouse and several railroad officials riding in the last car. Soon, as the train emerged from the Grant Hill tunnel, the locomotive engineer saw a horse-drawn wagon on the tracks, with the horses rearing in the air. The engineer immediately applied Westinghouse's experimental air brakes. The rapid stop threw all the passengers onto the floor, and the train came to a halt just four feet from the wagon. The railroad officials were impressed. The scheduled testing went on later that day, but the officials had already made up their minds. Shortly after that, Westinghouse established the Westinghouse Air Brake Co. and had no difficulty locating $500,000 in capitalization money. At 23 years of age, he was the president of a company.

Westinghouse's original air brake patent was the first of his 103 patents for railroad brakes. Just five years later, more than 7,000 passenger cars were equipped with his air brake. With his fertile imagination and driving energy, Westinghouse also developed a railroad signaling system that kept trains moving safely and efficiently. Moving on to natural gas systems, he devised a meter to measure gas consumption. Westinghouse also invented a leakproof piping system and an automatic cutoff that turned off the gas if pressure dropped below 0.25 psi. In all, he held 361 patents.

After the air brake, Westinghouse's most important contribution was probably his introduction of alternating current (ac) to the public. Westinghouse knew that ac was easily changed in transformers and could be more efficiently sent over high-voltage transmission wires. His company developed ac equipment and steam turbines to power the alternators. He bought Nikola Tesla's ac motor patents and hired Tesla to work on development. Thomas Edison, on the other hand, advocated the use of direct current (dc) because it could easily be stored in batteries. All the generating equipment Edison sold was dc. Because of their difference of opinion, Edison and Westinghouse became professional rivals competing for the huge consumer electrical market. For his part, Westinghouse avoided bringing personalities into the argument. Following a competition that newspapers called "The Battle of the Currents," Westinghouse won a crucial victory over Edison. He received the contract to install three massive 5,000 horsepower ac alternators at Niagara Falls. The alternators went on line on November 16, 1896, to deliver power to Buffalo, New York, 22 miles away.

Westinghouse received many honors and was the first American awarded Germany's Grashoff Medal. He became wealthy at a fairly early age, but in most ways he lived modestly. He had no expensive hobbies and did little traveling. However, he did have a large home on the east side of Pittsburgh for his wife and only child, and he had a private railroad car. Westinghouse was a private person with no vices beyond a tendency to eat too much. He kept no diary, wrote few letters, and refused to let reporters interview him. Few photographs were taken of him. Near the end of his life, he wrote the inscription for his own tombstone, which acknowledged the importance of his military service: "George Westinghouse (1846-1914)—Acting Third Assistant Engineer, U.S. Navy (1864-1865)."

References

Inventing: How the Masters Did It by Byron Vanderbilt, Moore Publishing, 1974.

George Westinghouse (1846-1994), Westinghouse Electric Corp., 1946.

American Science and Invention by Mitchell Wilson, Bonanza Books, 1960.

Dictionary of American Biography, Charles Scribner's Sons Publishers, 1932; with supplemental updates.

National Cyclopedia of American Biography, James T. White and Co. Publishers, 1891; with supplemental updates.

McGraw-Hill Encyclopedia of Biography, McGraw-Hill Publishers, 1973.

Asimov's Biographical Encyclopedia of Science and Technology by Isaac Asimov, Doubleday and Co. Publishers, 1964.

Thomas EDISON

Born: February 11, 1847, in Milan, Ohio

Died: October 18, 1931, in Orange, New Jersey

The words "Mary had a little lamb. . ." are far from technical. Nonetheless, they were the first intelligible sounds ever permanently recorded. Thomas Alva Edison said them loudly into the speaker of an experimental phonograph in 1877. Edison patented the phonograph—frequently described as his most novel and original invention—when he was a dark-haired and slender 30-year-old. It was his favorite invention and the 141st he patented out of a lifetime total of 1,093—the most ever granted by the U.S. Patent Office.

Called Al or Alva, Edison attended grade school for only a few weeks. His insatiable curiosity prompted him to ask many questions that his teachers could not answer. They were all happy when Edison's mother chose to teach him at home. At 14, Edison went to work for the Grand Trunk Railroad as a newsboy and general vendor. The job introduced him to steam engines, machine shops, telegraph offices, and other aspects of technology. Long train stops gave him time to visit Detroit's public library. There, he read books on science and technology that were not available in his hometown. The young Edison was a speed reader who went through books almost as fast as he could turn their pages. His native intelligence, nurtured by his mother's instruction, and his interest in technology combined to place him on a remarkable path of discovery and invention.

In 1869, Edison applied for a job in New York City. While he waited for his interview, a ticker-tape machine used to report stock prices broke down in a nearby office. Edison repaired the machine, and he received a job offer on the spot. He accepted the job, and

National Portrait Gallery, Smithsonian Institution

Thomas Edison at 23

in his spare time continued to experiment with the ticker-tape machine. He invented a better one and offered it to the president of the Gold and Stock Telegraph Co. Hoping to receive $5,000 for it, but willing to settle for $3,000, Edison was speechless when the president offered him $40,000. He was only 23. He used the money to establish the first industrial research laboratory. The photo on this page shows Edison that year. His stated goal was to produce a new invention every 10 days, and he didn't fall far short. He averaged one invention every 24 days for the rest of his life. Edison's first large laboratory was in Menlo Park, New Jersey, and the public commonly called him "the Wizard of Menlo Park."

Edison is probably best known for his 1879 electric lamp. However, he did not invent the incandescent light bulb. Sir Joseph Swan of England had one operating with a carbon filament almost 20 years earlier. Still, Edison made the lamp practical. One major problem was identifying an effective filament. Edison wanted one with a high melting point. It had to conduct electricity poorly, having a resistance value of about 100 ohms. The filament also had to be strong enough to cycle many times between room temperature and incandescent temperature. After three years of searching, Edison found that carbonized thread worked reasonably well. He used thread in his 1879 lamp, but he found that its fragile nature made it unsuitable for large-scale production. The first lamps he sold to the public used filaments made from carbonized bamboo fiber. They lasted an average of 600 hours, compared with 40 hours for carbonized cotton thread. Edison called the company that he established to manufacture his electric lamps the Edison General Electric Co.

Although Edison's laboratory typically employed upwards of 100 people, it was essentially a one-man operation. Edison's diversity, sense of detail, and powers of concentration seemed unequaled. Many of his inventions focused on electrical items, but he worked in other areas as well. For example, he received patents for a mimeograph duplicator, organic chemicals, iron production, and motion pictures, as well as the all-mechanical phonograph. Edison's imagination was boundless. By the time he reached his early thirties, he had already received half his lifetime total of patents.

Edison was an expert telegrapher, and he wanted to find a way to permanently record the dot-dash code. His first design used a sheet of waxed paper wrapped around a cylinder. A blunted needle connected to a telephone speaker rested on the paper. Edison rotated the cylinder while loudly making dots and dashes with a telegraph key. He then backed up the needle and tried to play back the sound through the telephone speaker. He couldn't tell a dot from a dash, but he noticed that the rapidly rotating cylinder produced a humming sound.

Edison's most trusted employee was John Kruesi, the man who made all of Edison's patent models. Edison asked Kruesi to make an experimental phonograph similar to the one in the photograph on this page. Edison wrapped the spiral-grooved brass cylinder with tin foil. A blunted needle attached to a speaker pressed against the tin foil. On December 3, 1877, Edison's squeaky voice shouted a nursery rhyme into the speaker as the cylinder was rotated by hand. Then, the blunted needle was returned to its starting point. As the cylinder was again rotated, the needle picked up peaks and valleys on the tin foil. A scratchy but intelligible sound came from the speaker. Edison was just as stunned as the others. He later said, "I was always afraid of things that worked the first time."

When the U.S. Patent Office made its usual search for an earlier patent, it found nothing that even remotely anticipated the phonograph. It was a completely new device. The phonograph was Edison's favorite invention, and the public loved it, too. Edison started to manufacture phonographs, and members of his staff developed an improved wax compound for the cylinders. To extend the life of the blunted needle, they developed an artificial jewel for its tip. Edison stayed with cylindrical records until 1912. Emile Berliner, a German-born American, patented the flat phonograph record in 1904. Flat-disk records were more compact and immediately began to replace cylinders. Edison eventually had to adopt the more practical disk records. In all, Edison manufactured phonographs for more than 40 years. Ironically, Edison himself was almost deaf from boyhood on.

Edison was strongly independent and

One of Edison's earliest phonographs

Smithsonian Institition Photo No. 9082

received countless medals during his life. He once joked that he had to measure them by the quart. After 13 years of marriage, his first wife died in 1884. He married again two years later. He had a total of six children. One son, Charles, became Secretary of the Navy in 1939 and Governor of New Jersey in 1941. Still, Edison wasn't much of a family man, and he spent most of his time at the laboratory. Work remained his greatest pleasure.

Edison had three close friends: naturalist John Burroughs, tire manufacturer Harvey Firestone, and automobile manufacturer Henry Ford. The four spent many vacations car camping in the Smoky Mountains and in other rustic settings. Starting in 1901, Edison regularly spent February and March at his winter home in Fort Myers, Florida. His friends often visited him there during the cold winter months, and Firestone planted a banyan tree that still grows on Edison's Florida grounds. For the 1929 golden anniversary of Edison's electric light, Ford moved Edison's Menlo Park laboratory to Greenfield Village, Ford's large public museum in Dearborn, Michigan.

Old age slowed Edison down but did not stop him. He received patents right up through the last year of his life. To pay their final respects to the world's most prolific inventor, tens of thousands of people waited in a cold New Jersey drizzle when Edison died.

References

A Streak of Luck—The Life and Legend of Thomas Alva Edison by Robert Conant, Seaview Books, 1979.

The Engineer by C. C. Furnas, Joe McCarthy, and the editors of Time Life Inc., 1966.

American Invention and Technology by Mitchell Wilson, Bonanza Books, 1960.

Dictionary of American Biography, Charles Scribner's Sons Publishers, 1932; with supplemental updates.

National Cyclopedia of American Biography, James T. White and Co. Publishers, 1891; with supplemental updates.

McGraw-Hill Encyclopedia of Biography, McGraw-Hill Publishers, 1973.

Asimov's Biographical Encyclopedia of Science and Technology by Isaac Asimov, Doubleday & Co. Publishers, 1964.

Alexander Graham BELL

Born:
March 3, 1847, in Edinburgh, Scotland

Died:
August 2, 1922, in Baddeck, Nova Scotia, Canada

The most valuable patent in United States history is number 174,465. It was issued in 1876 to a 29-year-old who had immigrated to America from Scotland. The patent was for the telephone. Today, there are almost 500 million telephones in use throughout the world, and Alexander Graham Bell's name will always be associated with the telephone.

Bell was born in a comfortable three-story house that we would now call a townhouse. The second of three sons, his parents named him Alexander after his grandfather, and he later adopted the name Graham after a family friend. Family members usually called him Graham. Both his father and grandfather were recognized professionals in the field of vocal physiology. Bell's father taught hearing-impaired people to speak, and he wrote several textbooks on the topic. He called his technique *visible speech*. Bell's grandfather was a professor of elocution at the University of London. Bell's mother was a sensitive musician and painter who helped him develop an appreciation for both hearing and music. He played music by ear and at one time considered a career in music.

During his early years, Bell's mother taught him at home. He eventually decided to follow his father's career path. He entered a Scottish academy at the age of 10, and by 20 he had received the finest university education available in Great Britain. Bell started working with his father in London and became his partner in 1869. In this context, he carried out experiments to determine how vowel sounds are produced. This started his interest in electricity.

National Portrait Gallery, Smithsonian Institution

Then, disaster uprooted his family. Bell's younger brother died of tuberculosis in 1867, then his older brother in 1870. Local physicians warned that Graham was showing signs of the disease and was at risk. His father had recently returned from the United States and Canada where he had been lecturing on speech articulation. He had found the weather there more pleasant than that in London, and he received greater professional recognition in America. Sacrificing a well-established career, the father moved the family to Canada. He chose to settle near Hamilton, Ontario, which he felt would provide a healthy climate for his surviving son. He was correct, and Graham soon fully recovered his health.

Replica of the world's first telephone. Bell spoke down into the large opening.

A principal at a Boston school for the hearing impaired requested a demonstration for her teachers on how to use visible speech. The younger Bell went in his father's place and spent three months at the school. He also visited other schools in the area. Due in part to his courteous manner, the demand for his services increased and he opened a school for teachers. Highly respected by his peers, Bell subsequently received an appointment as professor of vocal physiology at Boston University. He enjoyed life in Boston and met a young deaf woman whom he married in 1877. He became an American citizen on November 10, 1882.

Bell did some private teaching and was particularly successful with the son of Thomas Sanders, a wealthy leather merchant. To show his gratitude, Sanders offered to subsidize evening experiments Bell conducted in rooms he had rented on the top floor of a boarding house. Between 1873 and 1876, Bell worked along three related lines: making speech visible on paper (he called this making a *photoautograph*), a multiple telegraph, and an electric speaking telegraph. Bell found that he lacked the technical skill needed to make parts necessary for his experiments. He hired Thomas Watson to assist him. The two men became close friends, and Watson eventually received a share in the telephone patent as payment for his work.

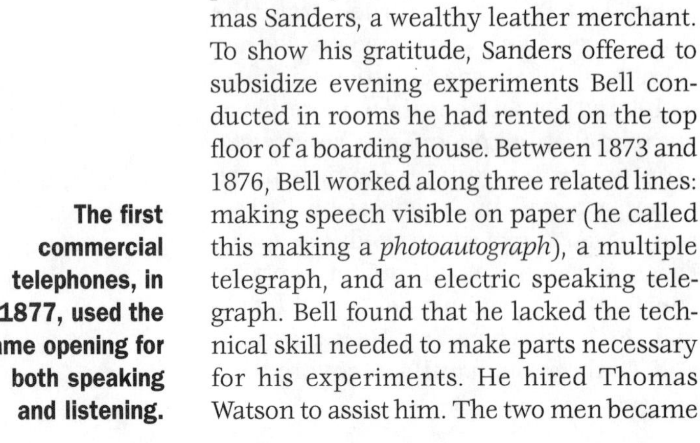

Bell said that the theory behind electrical transmission of the voice came to him while he was vacationing at his parents' home in Ontario in 1874. Putting the theory into practice presented a problem. But, Bell knew that Joseph Henry, first organizer of the Smithsonian Institution, had some knowledge in this area. Bell visited Henry in Washington, D.C., and Henry advised him that he was on the right track, that he should not yet publish his incomplete results, and that he should file a patent as soon as he could. Encouraged by the aging Henry, Bell worked even more diligently on the project. His patent application described a transmitter that had a short platinum wire attached directly to a diaphragm. As speech caused the diaphragm to vibrate, the wire moved up and down in a weak acid solution. When the wire went deeper, the resistance decreased. As the wire rose, the resistance increased. This meant that the sound controlled the current passing through the wire. The apparatus looked like a metal glass, set on a small metal frame, over a small coil of wire. On March 10, 1876, the telephone transmitted its first complete intelligible sentence. Bell shouted down into the mouthpiece of their experimental telephone: "Mr. Watson, come here. I want to see you." Watson was in another room but heard the sentences clearly.

Bell set up a display of his new telephone at the 1876 Centennial Exposition in Philadelphia. It failed to attract much attention until the Brazilian emperor, Dom Pedro II, requested a demonstration. Bell had set up schools for the hearing im-

The first commercial telephones, in 1877, used the same opening for both speaking and listening.

paired in Brazil and Dom Pedro recognized him. Dom Pedro listened at the receiver and exclaimed, "It talks!" Newspapers reported the event the following day, introducing the telephone to the public. Bell improved on his patented transmitter and receiver, and the Bell Telephone Co. came into existence in 1877. Passage of a year's time found 3,000 telephones in use and the first telephone exchange opened in New Haven, Connecticut. Alexander Graham Bell found himself rich and famous before he turned 30.

Bell's technical work in telephony ended completely in the early 1880s. He cheerfully admitted that keeping up with the rapidly changing technology did not appeal to him. However, many claimants came forward to contest his patent, and Bell became involved in about 600 court cases. The United States Supreme Court upheld all of his claims. The court declared Bell the discoverer of the only way that speech could be electrically transmitted.

The Scottish-born Bell had a summer home built on Cape Breton Island in Nova Scotia, which means "New Scotland." He maintained an interest in technology and used his remaining 45 years in many different ways. Bell financed the early stages of Albert Michelson's experiment to measure the speed of light. Michelson went on to become America's first scientific Nobel Prize winner in 1907. In 1883, Bell helped to establish and financed the first 12 years of *Science*, which is now a principal scientific journal. He served as president of the National Geographic Society from 1896 to 1904. He founded the Aerial Experiment Association with Glenn Curtiss and others. Bell's fertile mind helped produce the first airplane to make a public flight, in 1908, and the first to fly in the British Empire. He built a twin-engine hydrofoil boat that held the world's speed record for 10 years at 70.86 mph. He invented an iron lung and even investigated rocket propulsion. Bell considered himself a teacher first, and he continued to work with hearing-impaired people. Helen Keller, a close friend, dedicated her autobiography to him.

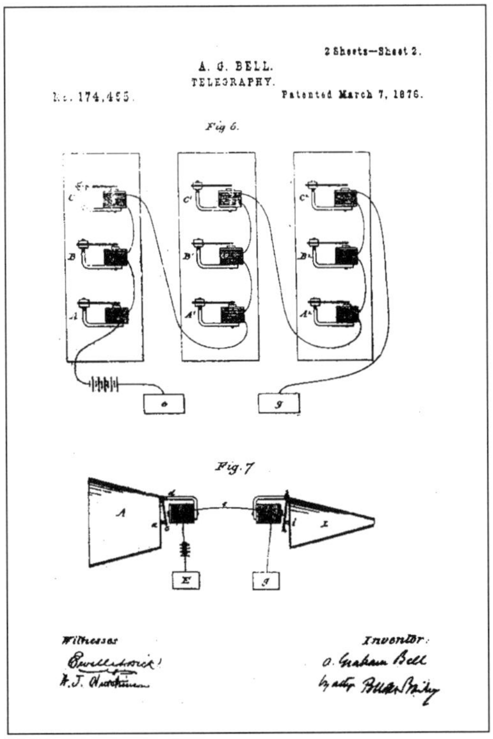

The first transcontinental telephone line was established in 1915 between New York and San Francisco. During opening ceremonies, Bell repeated his famous line to Thomas Watson on the West Coast: "Mr. Watson, come here. I want to see you." Watson humorously answered that he would be glad to come, but that it would take him a week to do so.

Bell often proudly proclaimed that he was an American by choice rather than by accident of birth. His grave stone in Nova Scotia reads: "Died a Citizen of the U.S.A."

References

Alexander Graham Bell, a pamphlet produced by AT&T, dated July 1979.

Those Inventive Americans, National Geographic Society, 1971.

Inventing: How the Masters Did It, by Byron M. Vanderbilt, Moore Publishing Co., 1974.

"Hello to History" by William Grimes, *New York Times News Service*, 7 March 1992.

Dictionary of American Biography, Charles Scribner's Sons Publishers, 1932; with supplemental updates.

National Cyclopedia of American Biography, James T. White & Co. Publishers, 1891; with supplemental updates.

McGraw-Hill Encyclopedia of Biography, McGraw-Hill Publishers, 1973.

Asimov's Biographical Encyclopedia of Science and Technology by Isaac Asimov, Doubleday & Co. Publishers, 1964.

Bell's patent for the telephone is the most valuable ever awarded. His later telephone patents were primarily refinements of his original one.

Lewis LATIMER

Born:
September 4, 1848, in Chelsea, Massachusetts

Died:
December 11, 1928, in Flushing, New York

Most people know that Thomas Edison made the incandescent lamp practical. However, many others contributed to improving the lamp to make it an affordable and durable consumer product. One such person was Lewis Latimer. He invented the first inexpensive and practical method for manufacturing carbon filaments and securing them to metal wires. Edison employed Latimer and so respected his opinion that Latimer often represented Edison in legal proceedings that took place in the early years of the electrical industry.

Latimer's father escaped from slavery in Virginia in the 1830s and settled in the Boston area. While Latimer was still a child, his father left the family. Latimer went to work at the age of 10 to help support his mother and three siblings. He had received little formal education. Latimer did the best he could to earn money through odd jobs and selling newspapers. After the Civil War broke out, he enlisted in the Union Navy at 16. He served on the side-wheeled steamer *U.S.S. Massasoit* and saw action on the James River in Virginia. After his honorable discharge in 1865, Latimer found employment with Crosby and Gould, patent attorneys in Boston. He had no related work experience, and he started out as an office boy. It was a fortunate turn of events because the job introduced him to inventions and inventing. In his spare time, Latimer taught himself drafting from library books and he practiced the principles with a used set of instruments. Noticing that he was a self-starter and willing to learn new skills, Crosby and Gould moved him up through the ranks. He was promoted to drafter and then chief drafter.

In his new position, Latimer prepared the drawings of inventions that went to the Patent Office in Washington. He worked in a building that was close to a school where Alexander Graham Bell conducted speech experiments. The two apparently became friends, and Bell used the offices of Crosby and Gould to patent the telephone. He specifically asked that Latimer prepare the drawings and descriptions for his 1876 telephone patent application.

During this period, Latimer received the first of his own several patents. The title of his 1874 patent was "Water Closets for Railroad Cars." This invention met with only limited success, but it launched Latimer on a career as an inventor.

Looking for a hardware-oriented position, Latimer found employment in 1880 with the United States Electric Lighting Co. in Bridgeport, Connecticut. The company's owner was Hiram Maxim, a prolific inventor whose best-known invention may have been the machine gun. Maxim also worked with gasoline engines and steam engines, and he had considerable experience with arc lighting. He wanted to extend his expertise to incandescent lighting and was emerging as Thomas Edison's most serious rival.

In his first significant project with Maxim, Latimer and co-worker Joseph Nichols investigated methods to attach metal conductors to fragile lightbulb filaments. The two developed a carbon filament with flattened ends that could be pinched with copper- or platinum-plated terminals. Copper wires connected the terminals to a metal base. With all components enclosed in a glass globe, the lightbulb worked like a modern incandescent lamp.

In developing his most important patent, Latimer conducted experiments that resulted in improved carbon filaments. Edison used bamboo strips that he curved and exposed to high heat in the absence of air. The result was a filament of almost pure carbon that glowed when electricity passed through it. Latimer's longer-lasting filaments came from textiles clamped under pressure between sheets of cardboard. The cardboard had grooves to hold the filament and flatten its ends to serve as electrical connections to the lamp base. Both the cardboard and the textile in the clamped assembly had the same rate of expansion. The assembly was placed in an oven and baked. The preshaped textile turned into a carbon filament that was stronger and longer lasting than Edison's strip of bamboo.

Latimer's methods proved so successful that Maxim had him set up an incandescent lamp department in Maxim's factory in Montreal. To communicate with the French-speaking workers, Latimer taught himself the language in record time. He found that he had to instruct the workers in all the processes for making Maxim lamps, including glass blowing. It took several months before the factory moved into full production. At that point, the public was still uncertain how

The filament in this lamp was manufactured using Lewis Latimer's patented technique.

to best use the new electric lamps, and Latimer's were primarily used to illuminate railroad stations.

After attending the Paris Exposition in 1881, Maxim never returned to the United States. Edison's display at the exposition completely overshadowed his, and Maxim decided to drop out of the lighting field. He settled in England, became a British citizen, was knighted by King Edward VII for his technical accomplishments, and died in 1916 at the age of 76.

Latimer moved to the Olmstead Electric Lighting Co. and then to the Acme Electric Light Co., both in the New York City area. He began his association with Edison in 1883 as a drafter and engineer in an office at 65 Fifth Avenue. A few years later, he began a series of frequent appearances as an expert witness on Edison's behalf before the U.S. Board of Patent Control. At the time, there was no national policy concerning the use of direct current (dc) or alternating current (ac) electricity. Edison had a great deal of dc-generating equipment, and he wanted the entire country to standardize on direct current. George Westinghouse, realizing the technical advantages of alternating current, just as strongly favored ac. Latimer testified on the issue to the Board of Patent Control, which attempted to establish standards.

Latimer also supervised the installation

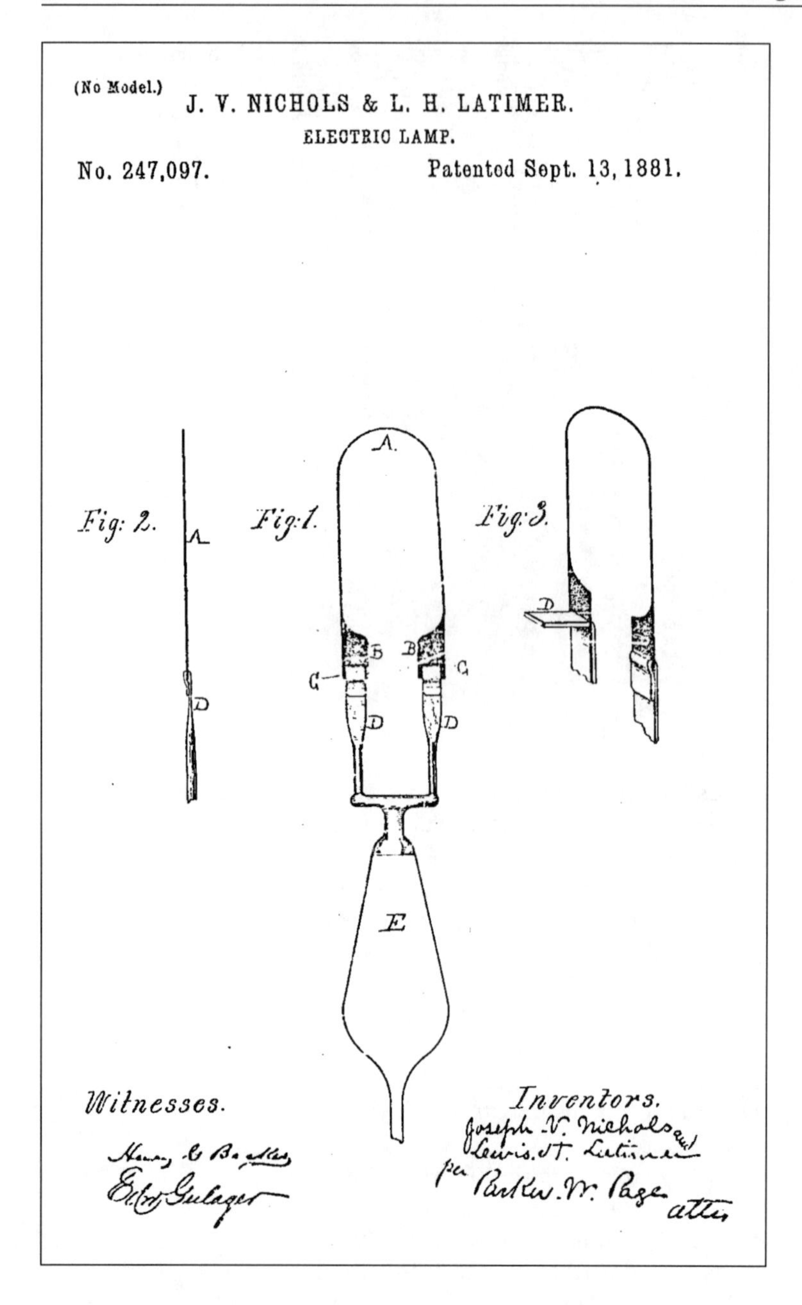

An early Latimer filament patent

of electric lamps in New York, Philadelphia, and London. He made a significant contribution to the infant industry by writing the first standard textbook on electric lighting: *Incandescent Electric Lighting*, published in 1890. He published it around the time that he transferred to Edison's seven-person legal department, where he served as chief drafter for patents under legal dispute. In 1896, Latimer left Edison to work as a drafter for the U.S. Board of Patent Control. He left that job in 1911 to become a full-time patent consultant. He retired in 1924.

Latimer was a member of the Edison Pioneers, a group of 28 technologists who had personally worked with Edison. Organized in 1918, the group held its annual meetings on February 11, Edison's birthday. An important year for the Pioneers was 1929, which marked the 50th anniversary of Edison's improvements to the incandescent lamp. It was called Light's Golden Jubilee, and the Edison Pioneers used the occasion to honor Latimer, who had died the previous year, along with Edison. The Pioneers recognized Latimer as the only black Edison Pioneer. Their official statement at the time of his death was: "Broad mindedness, versatility in the accomplishment of things intellectual and cultural, a linguist, a devoted husband and father, all were characteristics of him, and his genial presence will be missed from our gatherings."

Latimer was pleasant and well liked, and he was a man of many talents. Not only did he receive at least seven patents, but he was an amateur poet. On his 75th birthday, friends had his work printed and published in a 25-page booklet titled *Poems of Love and Life*. Latimer taught mechanical drawing to immigrants at the Henry Street Settlement in New York before illness claimed his life at the age of 80.

References

Black Pioneers of Science and Invention by Louis Haber, Harcourt Brace, 1970.

A Streak of Luck by Robert Conot, Seaview Books, 1979.

The Real McCoy: African-American Invention and Innovation by Portia P. James, Smithsonian Institution, 1989.

John FLEMING

Born:
November 29, 1849, in Lancaster, England

Died:
April 18, 1945, in Sidmouth, England

The late 1800s saw a rapid growth in desktop technologies like punched-card data processing, photography, and electricity. It wasn't obvious that electricity would soon have a major worldwide impact, but it was making surprising and diverse inroads. Thomas Edison (1847-1931) in America and Joseph Swan (1828-1914) in England manufactured incandescent lamps for sale in the 1880s. In 1888, Frank Sprague's (1857-1934) street cars in Richmond, Virginia, showed that electricity could be used for power applications. Telegraph systems by Charles Wheatstone (1802-1875) in 1837 and Samuel Morse (1791-1872) in 1840 showed the potential for its use in hard-wire communication. As the 19th century merged with the 20th, electricity's close relative electronics began to emerge. But only those who could interpret subtle experimental results could read the signs pointing toward electronics. Heinrich Hertz (1857-1894) first sent electronic signals through the air in Germany in 1886. But his experiment did not immediately result in wireless communication. No one knew how to control, or *modulate*, the signal to send readable information. That task faced John Fleming. He created the field of electronics with his 1904 invention of the first practical radio tube.

Born in northwest England, Fleming was the oldest child in his family. His family moved to London a few years later and Fleming spent the rest of his life there. His father was a minister. His mother came from a family that developed the manufacture of portland cement, which was named after the Isle of Portland off England's south coast. Fleming's family was comfortable, but not wealthy. Fleming received his primary edu-

The Science Museum/Science & Society Picture Library

cation in the local schools and graduated from the University College of London (UCL) in 1870. He had to work his way through college as a clerk. In spite of limited study time, Fleming tied for top honors with one other student. Fleming continued his education by winning scholarships to various colleges. His special interest was finding methods to measure electrical resistance. He developed a bridge circuit he called Fleming's Banjo. It was used for several years at university laboratories. Fleming's academic work was of high quality and he accepted a position as "demonstrator in mechanisms and applied mechanics" at Cambridge University in 1880.

Early in his career, Fleming moved be-

One of Fleming's original diodes. This is one of the special lamps that Fleming had made in 1889 for investigating the Edison effect. He modified it in 1904 to make his diode.

tween academic and industrial positions. Incandescent lamps had just gone into production and Fleming was fascinated by the prospects of an electrical world. He briefly taught at a college in Nottingham, then resigned to become a consultant for the Edison Electric Light Company in London. He stayed there for about three years. He was appointed professor of electrical technology at UCL in 1885 and remained there for 41 years. His early projects included transformers and high-voltage transmission, and he developed the right-hand rule. The rule relates the direction of the thumb, index finger, and middle finger to magnetic field, conductor motion, and electromotive force. Fleming worked with James Dewar (1842-1923) on electrical effects at temperatures near that of liquid air, about -330° Fahrenheit. He was comfortable in front of groups and had a reputation as an excellent teacher.

Fleming worked on improving the carbon filaments in incandescent lamps. They had a short life and tended to darken the glass globe with use. Edison had unsuccessfully tried to improve bulb performance by placing an extra electrode alongside the glowing filament. He hoped it would absorb excess carbon, somewhat like a "getter" in later radio tubes. Edison observed in 1883 that when the extra electrode was connected to a positive voltage, a small current flowed between it and the filament. This was later called the "Edison effect." Edison saw no special value in the effect but patented it anyway. (Edison seemed to patent everything in sight.) That phenomenon was the key to the first radio tube. Bright as Edison was, he did not see the subtle secret being revealed to him.

Undated production Fleming valve. Its label says, "Fleming Valve. J. Ambrose Fleming developed the vacuum tube diode. This was a practical application of the 'Edison Effect' in 1904."

Like some others, Fleming began serious research into the Edison effect in 1889. The merged Ediswan Electric Light Company gladly made 12 special lamps for his use. Many people had theories about the effect, but none had the answer. Fleming spent almost all his professional life in academics, but his experimental work made him the first in the field of applied electronics. He experimented with electrical conduction from glowing filaments in a vacuum. He hoped this would contribute to wireless communication, and he carried out countless experiments on transmission and reception. Fleming had a flair for linking complex electrical mathematics with their practical effect. He often made popular presentations on radio waves at the UCL, the Royal Institution, and the Royal Society. But the research seemed to lead nowhere and he left it for several years.

Fleming was a friend of Italian radio developer Guglielmo Marconi (1874-1937) and became a technical advisor to Marconi's company in 1899. He helped Marconi design a powerful wireless transmitter in southwestern England. That transmitter sent the first faint transatlantic radio signal to St. John's, Newfoundland, on December 12, 1901. It was the letter "S"—three dots—repeated over and over in Morse code.

The signal was hard to detect and Fleming

looked for ways to improve radio circuitry. He faced the problem of converting a weak alternating current (ac) into a direct current (dc) to operate a receiver. Fleming realized that his 1889 Edison-effect lamp could convert ac to dc because it let current flow in only one direction. He later wrote, "To my delight, I . . . found that we had, in this peculiar kind of electric lamp, a solution." With the idea in his head, Fleming easily made up a small circuit with one of the lamps. He experimentally confirmed in 1904 that his invention could detect radio waves. The electron had just been discovered and the Edison effect was known to be caused by electronic emission from a heated filament. Fleming's discovery is generally considered to be the beginning of the electronic era.

Fleming's diode had two electrodes inside a glass globe from which the air was evacuated. One electrode, the filament, was heated and emitted electrons. Incoming radio waves were directed toward the other electrode. It accepted electrons from the heated filament when it was positive. It repelled electrons when it was negative. In other words, it allowed current to flow in only one direction. Fleming called the electrode a *valve*, since it turned on when current flowed in one direction and turned off for flow in the other direction. It worked just like a one-way or check valve in hydraulics. The electronic symbol, a triangle, was borrowed from hydraulics. Britons still call the tubes "valves," but they are known as *vacuum tubes* in America. Fleming's invention allowed more precise detection of radio waves and it was heavily used in the early years of amplitude-modulated (AM) radio communication. It was an essential part of all transmitters and receivers for more than 50 years.

No. 803,684. PATENTED NOV. 7, 1905.
J. A. FLEMING.
INSTRUMENT FOR CONVERTING ALTERNATING ELECTRIC CURRENTS INTO CONTINUOUS CURRENTS.
APPLICATION FILED APR. 19, 1905.

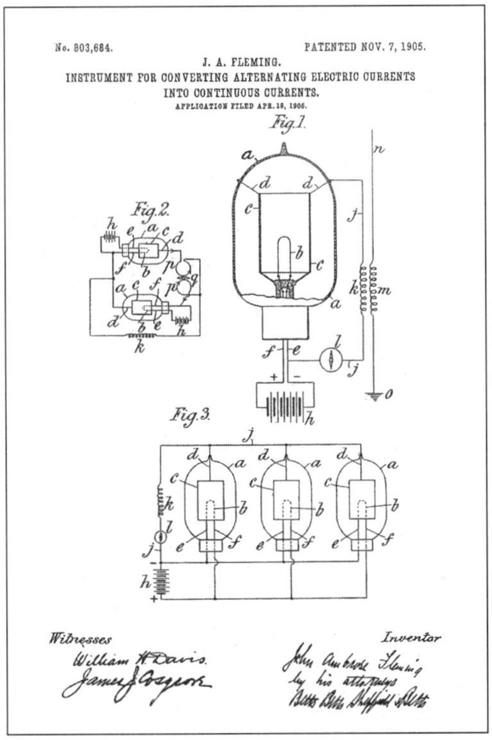

Fleming took out an American patent shortly after he applied for his British patent. Quite nicely drawn for the period, the hot carbon filament is identified as "b" and the secondary electrode as a metal cylinder "c."

Fleming was married to Clara Ripley for 30 years before her death in 1917. He married Olive Franks at the age of 84. He had no children. His favorite pastimes were wet-plate photography and painting with water colors. Like Edison, he was almost deaf and said that his poor hearing heightened his powers of concentration. Fleming was a captivating speaker. He made his last public presentation at the age of 90. He wrote three books and more than 100 technical papers. He was strong supporter of John Logie Baird (1888-1946) in his effort to establish television. Fleming served as president of the Television Society from 1930 until his death. He lived long enough to see his invention used in advanced electronic communication systems. A remarkably energetic individual, he remained professionally active almost all of his life. He died in 1945 at the age of 95.

Fleming's valve ushered in the field of practical electronic communication. It is still part of modern circuitry and used with television picture tubes, computer screens, and radio transmitters. But the valve was only a detector, not an amplifier. It laid the groundwork for the 1907 triode amplifier invented by Lee De Forest (1873-1961) in America. The triode made it possible to receive voice communication and opened up the world to anyone with a radio. The word *radio* was suggested as a trademark at a 1906 international convention in Berlin. It was widely adopted after 1915. No can say for sure when the Information Age began. A good argument can be made that it started with Fleming's simple electronic valve.

References

Making of the Modern World edited by Neil Cossons, John Murray Publishers, 1992.

The Principles of Wave Telegraphy and Telephony by John Ambrose Fleming, Longmans, Green, and Co. Publishers, 1910.

Edward WESTON

Born:
May 9, 1850, in Shropshire, England

Died:
August 20, 1936, in Montclair, New Jersey

Digital meters are so common today that we might not appreciate the measurement problems encountered by 19th-century electrical pioneers. All they had available were delicate galvanometers that made only comparative readings. They did not read in volts, amperes, or ohms. Galvanometers could not be moved during a test because the earth's magnetic field affected them. Thunderstorms and even the metal nails in shoes upset meter readings. The problem so exasperated Edward Weston that he took time from his busy life as a generator manufacturer to invent the world's first practical direct-reading voltmeter in 1886.

Weston was born in a brick farmhouse in a small town 150 miles northwest of London, England. His grandfather was a well-to-do farmer and his father was a respected mechanic. His family moved to the dynamic industrial city of Wolverhampton when Weston was 9. There, for the first time, he saw steel companies, foundries, and gas works. The school he attended introduced him to the exciting new field of electricity. He conducted textbook experiments at home using glass jars, sulfuric acid, and small pieces of copper and carbon. Yet his parents saw no future in his work with electricity, and they obtained a medical apprenticeship for him at 16. As an apprentice, Weston gained a knowledge of chemistry that would help him later in life, but he decided to leave the field of medicine after three years. While traveling by train to London, a chance meeting with a talkative and persuasive American tourist made him consider moving to the United States. Three weeks later, in May 1870, Weston spent practically all of his money for a steamship ticket to New York.

The Electrical World, June 1888

His chemistry background helped Weston secure employment with the American Nickel Plating Co. At the time, plating voltage came from clumsy and troublesome batteries that could produce only a thin and poorly bound layer of nickel. Weston developed a dynamo, or direct current generator, that produced higher and better-regulated electrical current than batteries could. This was the first time anyone had used a dynamo for electroplating, and the dynamo soon entirely replaced batteries in the process. The patent Weston took out at age 22 formed the basis for a successful electroplating partnership he established with one of his customers, Charles Theberath.

Income from this partnership allowed Weston to continue his work on dynamos, and he soon developed one for arc lighting that proved to be far superior to anything offered by his competitors. It included his invention of laminated pole pieces and cores that increased efficiency enormously—from 45 percent to 85 percent. Weston added other innovations to improve voltage regulation and soon opened the Weston Dynamo Electric Machine Co. in Newark, New Jersey. America's first commercial dynamo factory, it manufactured four bathtub-sized dynamos a day. Each sold for about $500. Weston held many patents related to arc lighting, such as the ones for the first copper-coated carbons that were universally used. He designed and sold complete lighting systems. His products were so highly regarded that Weston won the contract to illuminate the Brooklyn Bridge when it opened in 1883. The four lines of arc lamps that he suspended on what was then the highest structure in New York became something of a tourist attraction. They remained in place for 15 years, until incandescent lights replaced them.

Weston was also deeply involved with incandescent light production. He developed an improved pump for removing air from a glass globe. In 1885, he invented a way to coat weak carbon filaments with dense carbon to strengthen them and to provide a predictable resistance. He did this by sending high currents through filaments inside a container filled with natural gas. The heated filaments became coated with carbon from the natural gas. The process was called *hydrocarbon flashing*, and no successful carbon filament lamp was ever made without it.

Weston's decision to build a direct-reading voltmeter followed a week-long electrical experiment at the Franklin Institute's well-equipped generator laboratory in Philadelphia. He had expected the experiment to take only one day and he felt frustrated by the lack of proper meters. To that point, some inventors had unsuccessfully tried to make meters by using the heat generated from current flow. Others had tried to use magnetic effects by pulling two electromagnets together. Neither of these approaches worked well. Thomas Edison built the first electrical power plant in 1883 near Wall Street in New York. To make a crude ammeter he hung a nail from a string and positioned it near the electrical supply line. Different current flows caused magnetic effects that pulled the nail near the wire. Edison made adjustments based on this crude measurement.

Weston decided to expand on an 1881 French patent for an unsuccessful meter designed by Jacques Arsene D'Arsonval. The French inventor used a small coil of wire suspended by a filament inside a permanent magnetic field. A small amount of electricity sent to the coil converted the wire to an

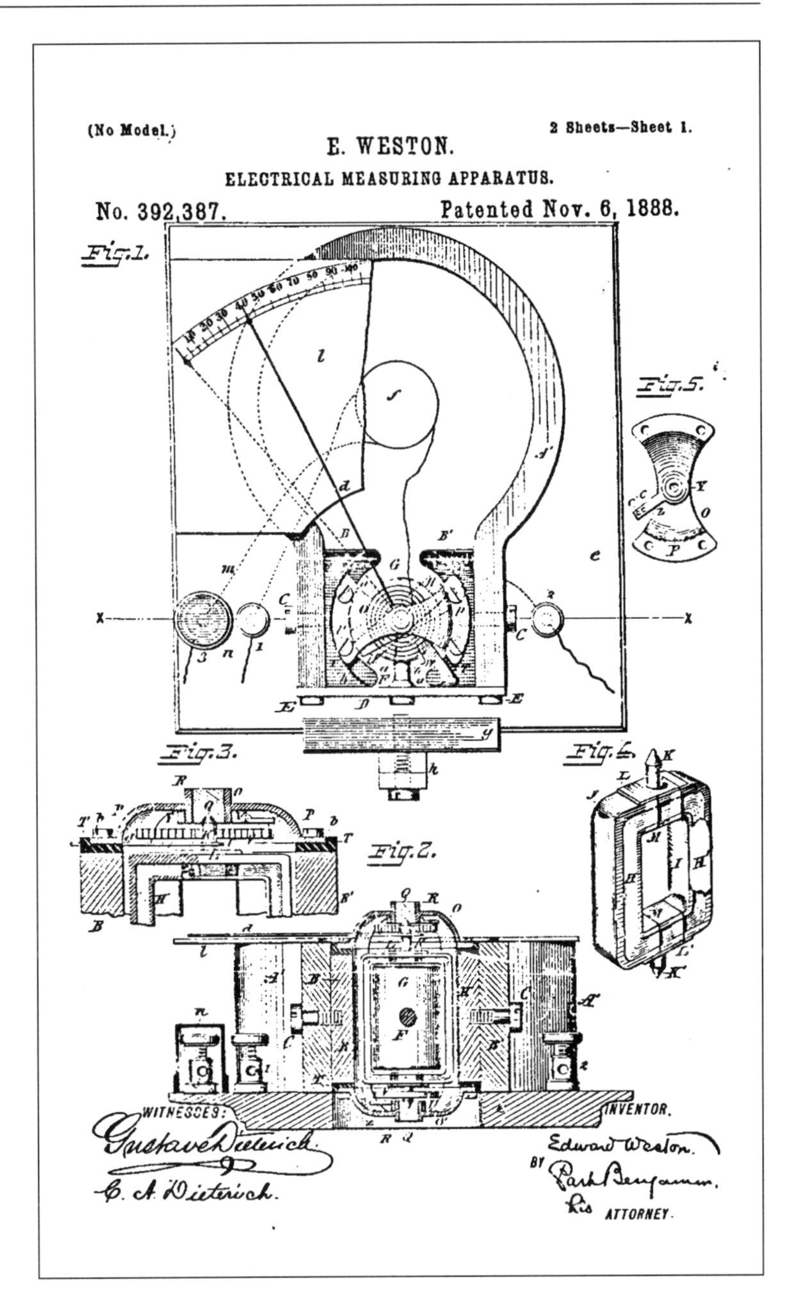

Weston's voltmeter patent used a D'Arsonval meter movement.

electromagnet. The two magnetic fields interacted, causing a slight rotation of the coil. A dial connected to the coil moved over a calibrated scale. The idea was good in theory, but D'Arsonval's specific design proved impractical. Nonetheless, the moving coil in electrical meters that have dial indicators is called a D'Arsonval meter movement.

Weston determined that the fundamental flaw of the French patent lay in the shape of the magnetic field surrounding the coil. The coil had to be completely encircled, and Weston was the first person to find a way to make permanent magnets in any shape. That was the key to a practical D'Arsonval meter movement. Weston spent two years perfecting his moving coil voltmeter before starting production in 1888. The 100 volt dc meter that he called Model One was the first portable voltmeter ever made. Weston carefully manufactured it with jeweled bearings, finely drawn coil wire, and nonmagnetic springs to provide a balance torque for the coil. The meter was an immediate success with schools, colleges, power companies, and electric equipment manufacturers. With the success of his voltmeter, Weston eventually saw his factory grow to 15 buildings on a 12-acre site.

Weston's catalog listed voltmeters with code names to avoid confusion when orders were sent by telegraph. He sold a 150 Vdc Reprint, a 150/300 Vdc Reptatus, and a 600 Vdc Requital. Each cost about $70 and was accurate to 1/4 percent. Weston invented the shunt resistor for ammeters, and his factory's production soon included ammeters, ohmmeters, and wattmeters, in addition to the voltmeters. He was the first in the electrical industry to use Bakelite for meter cases. Over the years, his meters took gold medals at seven international expositions.

Weston bought a large house in Newark for his wife and two sons. A two-story private laboratory he built in his backyard had modern tools for physical, chemical, and metallurgical research. Weston's residence was so complete that *Scientific American* described it as a "house of wonder." The recipient of 334 patents and a member of many professional organizations, Weston was a charter member of the American Institute of Electrical Engineers and later served as its president. The photograph on page 109 came from the 1888 announcement of his presidency, when Weston was 38. Weston also helped found the New Jersey Institute of Technology, and he received honorary doctorate degrees from McGill University in Montreal, Princeton University, Stevens Institute of Technology, and the University of Pennsylvania. He enjoyed collecting rare electrical books and autographs. Weston died of natural causes at the age of 86 while he cruised on his yacht in Long Island Sound after watching a boat race.

References

Measuring Invisibles, Weston Electrical Instrument Corp., 1938.

Edward Weston: Inventor—Scientist—Industrialist, Henry Berring, 1942.

Newark Evening News, 21 August 1936.

The Electrical World, 2 June 1888.

Weston Electrical Measuring Instruments—Catalog 15.

Trade journal information from Ford Museum Archives, Dearborn, MI.

Dictionary of American Biography, Charles Scribner's Sons Publishers, 1932; with supplemental updates.

National Cyclopedia of American Biography, James T. White and Co. Publishers, 1891; with supplemental updates.

Ottmar MERGANTHALER

It was so hard to print books in the 19th century that in the 1880s only 76 American public libraries had more than 300 volumes in their collection. Parents handed down school books to their children, and magazines were few, thin, and expensive. The largest daily newspapers were limited in size to eight pages because all type was set one letter at a time, exactly as it had been done for 400 years. A German immigrant's complex 1886 invention greatly speeded up the printing process. Ottmar Merganthaler's Linotype was such a remarkable machine that Thomas Edison called it the "eighth wonder of the world."

Merganthaler's father was a village schoolmaster in Germany and his mother was the daughter of a schoolmaster. He was brought up in an academic setting, but his family had a limited income. Like many 19th-century technologists, Mergan–thaler showed an early interest in mechanical devices. Before he had entered his teens, he had repaired the town clock after the local clockmaker gave up on the task. Merganthaler wanted to study engineering or science, but his parents wanted him to teach school. He resisted the idea and reached a compromise by agreeing to take a clockmaker apprenticeship as a 14-year-old. He also attended night classes to expand his technical knowledge. Mergan-thaler considered his work on clocks the most useful work he ever performed. He said, "Above all, watchmaking taught me precision. . . . I realized that if a [watch] was to work, it must be considered as a whole, that each part had to be perfect in itself and also harmonize with every other." That experience helped him later in life, when he spent

Smithsonian Institution Photo No. 32896

Born:
May 11, 1854, in Hatchel, Germany

Died:
October 28, 1899, in Baltimore, Maryland

years working on the fiendishly complex Linotype.

The Franco-Prussian War of 1870-71 encouraged many Germans to leave their homeland. Following completion of his apprenticeship, Merganthaler was among the 125,000 who emigrated in 1872. A cousin of his, August Hahl, who lived in Washington, D.C., offered him a job. The cousin even advanced Merganthaler the necessary travel money.

The goal of all inventors is to obtain a patent, and in the 1870s a model was required for patent applications. Washington, D.C., was the center of the patent model industry and Hahl owned a machine shop that specialized in making models. He also

This 1915 Linotype is quite similar to the one Merganthaler invented in 1886.

made measuring instruments for the emerging field of electrical technology. Because of his clockmaking experience, Merganthaler adapted well to his new job. He made rapid progress in his new profession. Hahl even kept him on the payroll when a downturn in the national economy mandated a move to Baltimore.

In August 1876, shortly before Merganthaler became an American citizen, an inventor approached him with the plans for a kind of typewriter. The machine used lithographic techniques to transfer an entire page of text. Constructing the model introduced Merganthaler to the printing industry's fruitless efforts to develop an automatic typesetting machine. More than 200 inventors over a period of 50 years had tried to patent a typesetter. The author Mark Twain went bankrupt in the late 1880s while financing one designed by James Paige that had 18,000 parts. Merganthaler began to work on an automatic typesetter in his spare time after hearing of a reward. The New York City newspapers offered $500,000 for a machine that would save at least 25 percent of the work of hand composition.

Merganthaler and Hahl began a partnership in 1878 that lasted five years. Then, Merganthaler decided to establish his own machine shop. There was plenty of patent and instrument work to go around, and Merganthaler's business prospered. He had not lost interest in inventing a printing machine and he spent every spare minute pursuing the idea. Merganthaler felt challenged by the machine's mechanical complexity. He had an unusual advantage over others with the same objective: Merganthaler had no background in printing and could approach the problem from a fresh perspective. Where others tried to duplicate the hand motions of a human typesetter, Merganthaler focused on molds. He worked on a technique for casting an entire line of type at once.

Merganthaler's first important invention came in 1883. He called it the First Band Machine. About the size of a large refrigerator, the machine had a series of vertical bars, with the alphabet and other special characters on each bar. When the operator typed one letter, a bar would descend to bring the character to a certain level. After a line of type was assembled, a papier-mache strip was pushed into the line. It produced a one-line mold, or matrix. When all the mold strips for a single page were assembled in a frame, molten metal was poured into all of them at once. This produced a completed printing plate.

Unfortunately, the machine was not a commercial success, and Merganthaler made only a few of them. To pay for its high cost, the rate of composition had to be several times faster than what people could do by hand, and the First Band Machine wasn't that fast. Merganthaler's financial backers grew concerned that he had greater interest in producing a perfect machine than in producing a workable one. It was a valid assessment on their part because Merganthaler was a perfectionist with what was typically considered a German's eye for precision. He had abandoned friends when he thought their ideas were wrong, and he occasionally went into debt when he neglected the responsibilities of his business.

July 26, 1884, is often cited as the birthday of the Linotype machine. On that date, Merganthaler demonstrated his Second Band Machine to a small group of potential purchasers. Its major improvement was the elimination of the papier-mache operation. Long thin brass bars with individual type molds dropped into alignment with others at the touch of a letter on the 90-key keyboard. Hot metal pressed against the line to form a slug, and several slugs combined to make a page. The operation was not perfect—people still had to perform right justification by hand, for example, and the machine was quite expensive. However, Merganthaler's financial backers were encouraged by the results and provided the money for the first production Linotype. On July 3, 1886, a portion of the *New York Tribune* newspaper was set with a Merganthaler experimental typesetter. Whitelaw Reid, the publisher, saw the machine at work and exclaimed, "Ottmar, you've done it . . . a line o' type." That was the casual christening of the most potent machine of its age. The news-paper purchased 12 of the new machines, and before long the first hundred were in use.

A printing boom soon began. More people were hired at higher wages for shorter hours as newspapers increased in number and size. Cost dropped from 3¢ for an eight-page newspaper in pre-Linotype days, to 1¢ or 2¢ for many more pages after Merganthaler's invention. Within 20 years, daily newspaper circulation in America had increased from 3.6 million to 33 million. Because of Merganthaler's invention, the magazine industry emerged, and schools all over the nation could buy inexpensive books.The illiteracy rate dropped from 17 percent to 5 percent. By 1900, there were over 8,000 Linotypes operating throughout the world.

Merganthaler became wealthy, but he never lost interest in his invention. He devised more than 50 patented improvements over the next few years. He received awards from the Franklin Institute in Philadelphia and Cooper Union in New York. Constant work and anxiety had undermined Merganthaler's health, though, and he contracted tuberculosis while in his forties. He moved to Deming, New Mexico, in hope of improving his health. There, an unfortunate fire in 1897 destroyed his home, technical papers, and an autobiography he had started. Against the advice of physicians, Merganthaler returned to Baltimore where he died at the age of 45.

References

Mechanisms of the Linotype and Intertype by Oscar Abel and Windsor Straw, Brookings Lebarwarts Press, 1961.

"Merganthaler's Wonderful Machine" by Michael Scully, *The Reader's Digest,* March 1953.

Publications provided by Merganthaler Linotype Co.

Dictionary of American Biography, Charles Scribner's Sons Publishers, 1932; with supplemental updates.

National Cyclopedia of American Biography, James T. White and Co. Publishers, 1891; with supplemental updates.

McGraw-Hill Encyclopedia of Biography, McGraw-Hill Publishers, 1973.

One of Merganthaler's early patents for making a matrix, or group of lettters

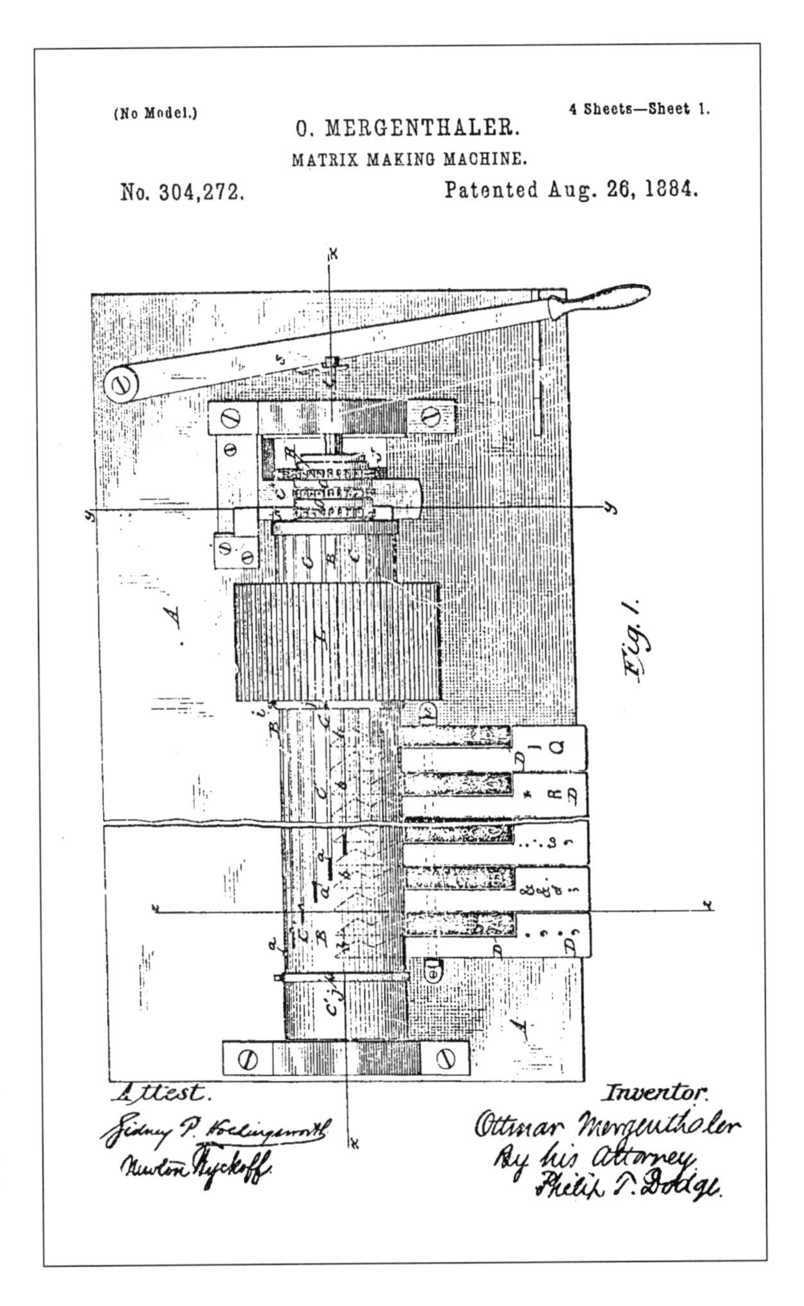

George EASTMAN

Born:
July 12, 1854, in Waterville, New York

Died:
March 14, 1932, in Rochester, New York

The small hand-held cameras that most amateurs use for taking personal photographs did not originate in the 20th century. George Eastman introduced the world's first conveniently sized, mass-produced camera intended for use by the public in June 1888. It came loaded with 100 exposures. After taking all the pictures, the purchaser returned the entire camera to Eastman's factory. For $10, the photographs were developed and printed, and a fresh roll of film was inserted. Eastman marketed his camera as the No. 1 Kodak.

Eastman went to work at age 14 to support his widowed mother and two sisters, when his father died unexpectedly. His first job as a messenger boy for an insurance company paid just $3 per week. Eastman later took work as a junior clerk in a bank in Rochester, New York.

When he was 24, Eastman took his first vacation, a photographic trip to Mackinac Island, Michigan. At the time, the simplest photographic process available used wet plates of glass. A chemically cleaned glass plate was coated with a sticky substance called a collodion. Next came a wet, light-sensitive silver salt emulsion. The operation had to be carried out in a completely dark tent. Using a special light-tight holder, the photographer placed the wet plate in a large camera. The exposed plate had to be developed before it dried. For $94, Eastman bought the necessary equipment, which included a large tripod-mounted camera and a dark tent. Glass plates, chemicals, tanks, several plate holders, and some miscellaneous items rounded out his supplies. Learning to use the equipment cost him $5.

Courtesy of Eastman Kodak Co.

Formal portrait of George Eastman, taken around 1884

Eastman became very absorbed in photography and wanted to simplify the complicated process. Particularly interested in making a dry emulsion, he spent evenings conducting experiments in his kitchen. Most earlier investigations of dry emulsions were done in England, and to learn about them Eastman read every issue of the *British Journal of Photography* that he could obtain. Forty years had passed since the French inventor Louis Daguerre introduced his daguerreotype technique for recording images. However, no U.S. citizen had ever made a significant contribution to the technology of photography.

Initially, Eastman just wanted to simplify picture taking for his own enjoyment. However, he soon considered the possibilities of producing dry plates for the professional market. He worked tirelessly during his free time. Eastman dreaded the potential poverty he felt could overtake his family, as had occurred when his father died. He invented a process for mass producing dry photographic plates in 1879. Traveling overseas with drawings and specifications, he obtained his first patent in England. The sale of his patent rights for $2,500 provided enough money to establish a U.S. factory, and Eastman left his job at the bank. His dry-plate business soon soared.

While looking for a less-heavy and more flexible support for his emulsion, Eastman experimented with paper. Professional photographers did not like the product. The paper's grain was reproduced in the print after removal of the emulsion from the paper during development. Eastman then decided to test the amateur market. He invented a camera the public could use, becoming the first manufacturer to do so. With the single venture of the No. 1 Kodak, Eastman brought amateur photography into being.

Eastman was the first person to practice the modern techniques of large-scale production at low cost for a world market. The Kodak was loaded with a long roll of flexible film. A popular slogan of the time that Eastman created was: "You push the button, we do the rest." Such an idea was unheard of at a time when average people thought photography was an advanced hobby only for those who had considerable knowledge of chemistry. The public loved the new camera and made Eastman an extraordinarily wealthy man.

Professional photographers did not take kindly to amateur photography. Eastman did not feel particularly concerned, and he always thought of himself as an amateur who took a cumbersome procedure and simplified it. By the mid-1890s, his cameras and emulsion development service had expanded several times but still could not keep up with the demand.

Eastman made up the word "Kodak" as a trade name for his camera. He once explained, "I devised the name myself. The letter K had been a favorite with me. It seemed a strong, incisive sort of letter. It became a question of trying out a great number of combinations of letters that made words starting and ending with K. The word Kodak was the result." The Eastman Kodak Co. was incorporated in 1892 with a capitalization of $35 million.

Eastman was one of the first commercial users of celluloid as an emulsion backing. Celluloid was invented by the American John Wesley Hyatt in 1869. Eastman's photographic film—for the first time a photographic medium could properly be called "film"—was manufactured by spreading a celluloid solution on a 200'-long by 42"-wide glass table. The smooth transparent celluloid remained with the emulsion after developing, eliminating the delicate emulsion-stripping operation. This flexible, strong, and transparent film made motion pictures possible. Its development was so important to the film industry that the word "celluloid" has been used to describe motion pictures.

Eastman always felt the need to share his good fortune in ways that reflected his social views. Most of his monetary gifts went to colleges, particularly ones with dental clinics. He wanted children to benefit from early dental care so they could have "a better chance in life with better looks, better health, and more vigor." Because his best technical assistants came from the Massachusetts Institute of Technology (MIT), he anonymously gave a total of $20 million to that school. On just one day in 1924, he gave away a total of

This first-ever amateur camera was named the No. 1 Kodak. Casually called the Kodak, 100,000 were manufactured between 1888 and 1896.

Courtesy of Eastman Kodak Co.

$30 million to the Hampton Institute, MIT, the Tuskegee Institute, and the University of Rochester. Dental clinics in Brussels, London, Paris, Rome, and Stockholm also received his financial support. In all, he gave away his entire personal fortune, estimated at between $75 and $100 million.

Eastman extended his generosity to his employees by declaring a wage dividend in 1912. It amounted to 2 percent on all wages received over the previous five years. He developed other employee benefit programs such as a medical department, high-quality lunchroom facilities, shorter hours, and the sale of company stock at reduced rates. By the 1920s—long before employers typically offered such benefits—Eastman had established a retirement plan, life insurance, and disability benefits for his employees.

Eastman succeeded to such a degree in promoting responsible managers that by 1932 his direction was no longer required at the Eastman Kodak Co. Also, he could not understand the details of his company's chemical research into such advanced products as color film and improved amateur cameras. A lifelong bachelor, his large home had not had a first lady since his mother's death 25 years earlier. Eastman felt he was too old and ill to carry on. He wrote on a piece of paper, "My work is done, why wait?" before taking his own life in 1932.

References

Technology in America by Carroll W. Pursell, Jr., MIT Press, 1983.

American Science and Invention by Mitchell Wilson, Bonanza Books, 1960.

"George Eastman" by O. N. Solbert, in *Image—The Journal of Photography of the George Eastman House*, November 1953.

A Brief History, a pamphlet from the Eastman Kodak Co., Rochester, New York.

Dictionary of American Biography, Charles Scribner's Sons Publishers, 1932; with supplemental updates.

National Cyclopedia of American Biography, James T. White and Co. Publishers, 1891; with supplemental updates.

McGraw-Hill Encyclopedia of Biography, McGraw-Hill Publishers, 1973.

Asimov's Biographical Encyclopedia of Science and Technology, by Isaac Asimov, Doubleday and Co. Publishers, 1964.

Albert Blake DICK

Born:
April 16, 1856, in Galesburg, Illinois

Died:
August 15, 1934, in Chicago, Illinois

If you say the word *mimeograph* to a teacher, the teacher probably will think of a motor-operated machine that rapidly makes many copies of classroom materials. However, the first mimeograph machine made just one copy at a time in a manner similar to silk screening. Albert Blake Dick invented the first successful mimeograph in 1887.

Dick received a typical elementary and high school education. In his first job out of school, he worked for a manufacturer of agricultural equipment in his hometown. He stayed there for five years, beginning to establish himself in the business world. Dick also worked for John Deere in Moline, Illinois, and finally wound up in Chicago, in his late twenties. There, he established a successful lumber company.

To communicate with mills and lumber yards, Dick developed the idea of sending a daily inquiry sheet. Responses to the sheet gave him up-to-date information about where he could obtain needed lumber sizes. He would typically send out 50 or more identical, hand-written letters. Although Christopher Sholes had invented a practical typewriter in 1868, American business people still relied heavily on handwriting as their basic communication method. Dick wanted to find a simple way to make multiple copies in an office environment.

Dick tried several ideas, but his successful solution followed a casual experiment he conducted at his desk. He had just eaten a piece of candy wrapped in waxed paper. He placed the paper over a file and pulled a nail over the paper. Dick held the paper up and saw that it was perforated along the line made by the nail. He reasoned that if he could force heavy ink through the holes and

Photo courtesy of A.B. Dick Co.

onto a sheet of paper, he could duplicate the line. At this early stage, he only considered the duplication of handwriting.

Dick still needed to develop a suitable file-like plate and to make a durable wax stencil master. After solving these problems, he was ready to apply for a patent. In conducting the necessary patent search, he discovered that Thomas Edison had tackled a similar problem. Edison had an existing patent for a vibrating electric pen that made a series of holes in a wax master. The pen had a small

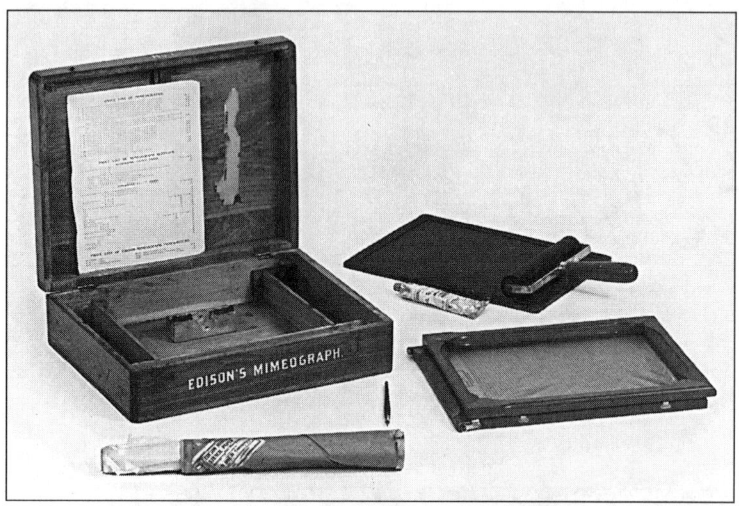

Photo courtesy of A.B. Dick Co.

Early mimeograph duplicating hardware and supplies

low-voltage dc motor at its top that caused its pencil-sized pointed steel shaft to vibrate 8,000 times a minute. A person writing with the device would make a series of holes in a wax master, much the same as in Dick's method. Ink was rolled through the master onto a sheet of paper to make the copy. While Edison's invention was only moderately successful, it is particularly noteworthy because it was the first product with an electric motor to go into production.

Edison and Dick met, liked each other from the start, and became lifelong friends. They had no trouble developing an agreement on the mimeograph. Edison provided some initial financing and invented a device for making the wax masters. Although the use of Dick's flat-bed, file-like surface was essential to the invention's success, he emphasized Edison's name on the mimeograph label. Labels of the early models read: "The Edison Mimeograph, Originally Designed and Patented by Mr. Thomas A. Edison, Made By A. B. Dick Company, Chicago U.S.A." The word *mimeograph* originated in the Greek words *mime* (to copy) and *graph* (to draw).

To use the mimeograph, an office employee placed a stiff wax master over a fine file-like plate, or sheet, of specially frosted glass. The person then hand wrote a message on the master with a metal-tipped stylus. Each written line or word created many holes in the master. The holes converted the wax master to a stencil. The stencil was lifted from the plate and placed on a sheet of paper. The employee rolled an inked, hard-rubber roller over the stencil. That forced ink through the many small holes and made a single copy. The process was repeated for as many copies as necessary. Each stencil could make up to several hundred copies.

Dick aggressively marketed the Edison-Dick mimeograph, as the product was originally called. His company made mimeographs in an eight-story building on the corner of Lake and LaSalle Streets in Chicago. Within five years, the company had 1,700 employees.

The original mimeograph used a handwriting technique instead of a typewriter for two reasons. Early masters were stiff and could not bend to fit into a typewriter. Also, they were not strong enough to withstand the impact of the keys. Experimental masters shredded during typewriter tests.

Dick accidentally discovered a stronger master material while buying a pair of shoes. The shoes came wrapped in soft, long-fiber paper. Dick thought that such paper might make a suitable carrier for the wax on the master. Dick's idea was similar to the one George Eastman had when he used paper in 1884 as a carrier for light-sensitive emulsion in cameras used by amateur photographers. Dick experimented with different papers and ultimately developed his own made from the fibers of a particular species of hazel bushes. The bushes grew only in high mountains on islands off the coast of Japan. Dick coated the paper with an improved wax made from seven parts paraffin to one part ordinary lard. Both sides were sealed with a thin layer of elastic varnish. The sandwiched combination was flexible and yet strong enough to resist the impact of typewriter type. Typing compressed the special wax away from the paper, exposing clean holes that allowed ink to pass through freely. Typewriters became

Label on an early mimeograph

more popular in businesses and schools, and the mimeograph offered rapid duplication.

Hand-rolled copies could be made at the rate of only a few copies per minute. It wasn't until 1900 that Dick developed a rotary mimeograph machine. The frame of the flat bed was bent into the shape of a half cylinder, similar to that found in modern mimeograph machines. Ink passed through the stencil from inside the cylinder. The hand-cranked device did not revolve continuously. It rocked back and forth as each sheet of paper was fed into the machine. Four years later, Dick introduced the Model 75, the first completely rotary-type mimeograph machine.

Dick served as a trustee of Lake Forest College, was a board member of several banks, and participated in many civic activities. Widowed in 1885, he married a second time. He had a total of five children. He named one of his sons Edison, in honor of Thomas Edison. Although Dick held many patents associated with mimeographic duplications, he often referred to Edison as the originator of stencil duplication. Dick was the head of his company for 51 years. Few organizations can duplicate that record for continuous leadership.

References

Materials from the A. B. Dick Co., Chicago, IL, circa 1934 and 1949.

National Cyclopedia of American Biography, James T. White and Co. Publishers, 1891; with supplemental updates.

Granville WOODS

Born:
April 23, 1856, in Columbus, Ohio

Died:
January 30, 1910, in New York, New York

Black Americans have made significant contributions to our society while going about their professional lives. The U.S. Patent Office generally did not keep records of an inventor's race, yet through 1900, at least 104 black Americans received more than 375 patents. Many of their inventions were as important as those of better-known technologists. Yet, if past inventors of all races are generally forgotten today, this is especially true of black inventors.

Black American technologists have always been innovators and discoverers. Their contributions are all the more remarkable considering the discouraging social and political barriers they frequently encountered. We know little about many of them because they received only limited publicity. Very few wrote biographies, and detailed information about their lives is scattered. No better example of this can be offered than the sketchy records of the life of Granville Woods, a genius in the field of electricity.

Woods was born in central Ohio, and he attended elementary school for a few brief years. He started working in a machine shop at the age of 10, and he did his best to educate himself in the evening through private tutoring. A voracious reader throughout his life, he was, for the most part, self-taught. At age 13, he learned that metal rails connected the Atlantic and Pacific Oceans, and he read everything he could find about railroads. He left home at 16 and moved to Missouri to work first as a fireman and then as a locomotive engineer for the Iron Mountain Railroad. He enjoyed all aspects of railroad work and developed an interest in telegraphy. Telegraph lines followed the rail lines, and Woods learned all he could about electricity and electrical transmission of information.

Woods held many different jobs in his late teens and twenties. From his work with the railroad, he moved to a steel mill in Springfield, Illinois, and then to a machine shop in New York City. Just before he turned 22, he signed up for a four-year tour aboard the British steamer *Ironsides*. Between 1882 and 1884, Woods worked for the Danville and Southern Railroad. During the evening, he worked on new ideas related to his work, and he received his first patent in early 1884. It was for an improved steam boiler furnace that used less coal than conventional boilers did. Receiving the patent encouraged

Woods to strike out on his own. Throughout his working life, he had saved part of his wages, and in 1884 he had enough money to establish a business. He and his brother Lyates opened a machine shop in Cincinnati. They called it the Woods Electric Co.

It was a small company that produced specialty items for its customers. The Woods brothers made such things as gears, generators, and tools. They repaired pumps, measuring instruments, and household items. Woods worked on electrical devices and took out his second patent 1884. It was for a highly efficient telephone transmitter that used a flexible diaphragm and carbon particles. A person's voice caused the diaphragm to vibrate and press against a small box of carbon particles. When the particles came close together, they conducted electricity quite well. When they spread apart, they conducted electricity only weakly. The action produced a series of strong and weak electrical currents that followed the sound patterns of a person's voice. This method made the telephone more practical. Woods sold the patent rights to the American Bell Telephone Co. of Boston, without a requirement that his name be used with the transmitter. That decision was probably a mistake, but it was a pattern Woods always followed. It is the main reason that Woods is not well remembered.

Selling patents provided Woods with money that supported further experiments at the machine shop. Some of his customers included the General Electric Co., Westinghouse Electric and Manufacturing Co., and the American Engineering Co.

Most of Woods's work was in the field of electricity. This was an emerging technology that tended to attract the brightest and most capable technologists. Woods's best year was 1887, when he received seven patents. He designed a regulator for electric motors that permitted the user to vary the speed without using energy-robbing resistors. The regulator proved so effective that it produced an energy savings of 40 percent. Woods's most sophisticated invention was the induction telegraph, a system that allowed communication between a moving train and a railroad station. It worked by laying an electrical wire on the ground between the rails. Another wire was suspended underneath a railroad car that carried telegraph equipment. The second wire hung 10" above the first and parallel to it. The moving train transmitted telegraph messages by sending electrical impulses to the wire suspended from the car. Those impulses induced a current in the receiving wire between the rails. The telegrapher at the railroad station picked up the dots and dashes of the induced current. The technique permitted both sending and receiving messages on a moving train.

Loops of wire attached to a moving railroad car picked up telegraph signals from a wire between the rails.

Apparently influenced by the success of Thomas Edison, Woods reorganized his company to manufacture his own inventions. However, he did not have the proper tool

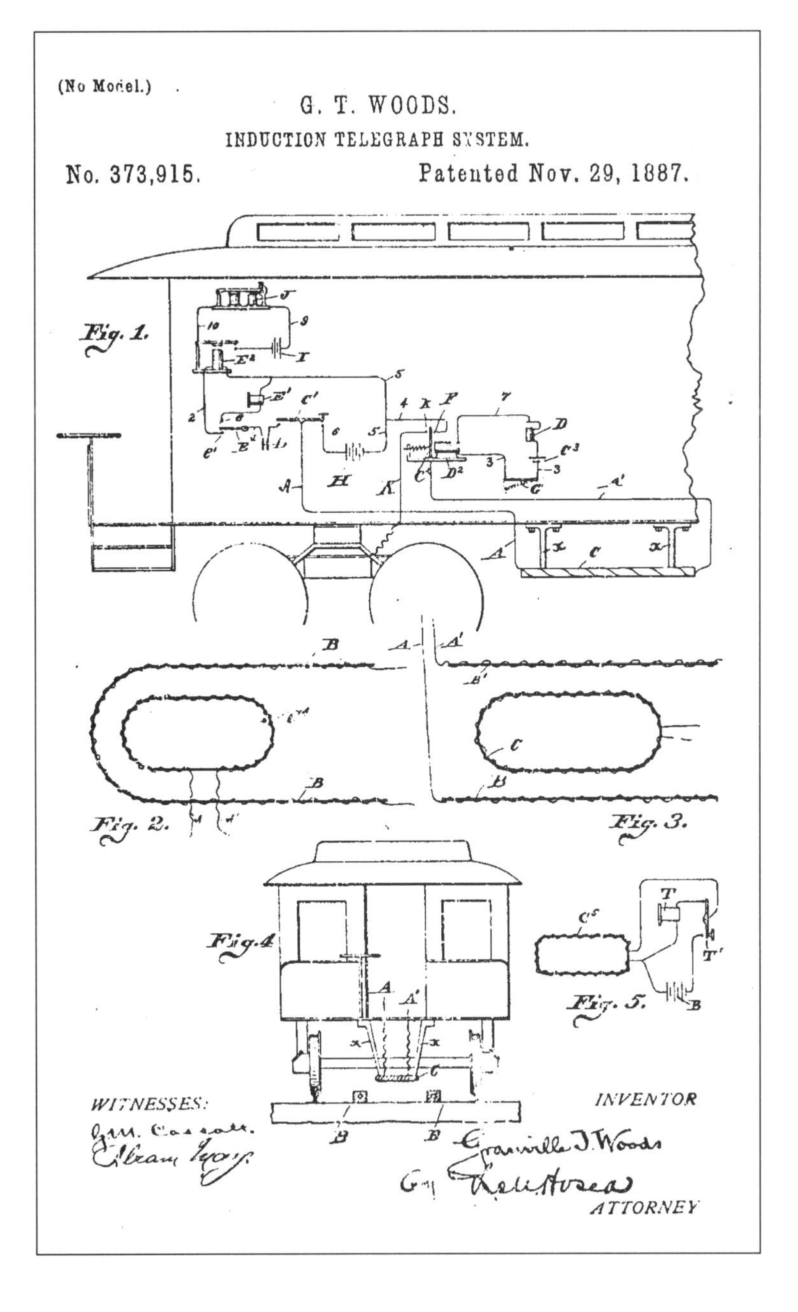

ing and went back to selling his patents. Woods wanted to make electricity practical for transportation, and many of his approximately 50 patents reflect that interest. In 1888, he worked on a system of electrically driven street cars for mass transportation in cities. The cars drew their power from an overhead line. Using electricity eliminated the clouds of smoke and soot from steam engines operating in cities. One of Woods's inventions that continues to modern times is the third-rail power pickup on subways and urban commuter trains. Subways draw electricity from a protected power rail outside the two original tracks. Woods received a patent for it in 1901. He received patents for electromagnetically controlled air brakes on trains and automatic circuit breakers for operator safety. Having an expansive mind, Woods also patented an egg incubator, a small train for use in amusement parks, and improved techniques for constructing tunnels. Around 1890, he sold his share of the business and moved to New York City.

Shortly before his death, Woods was practically penniless, but the reason is unclear. He had succeeded as an inventor and machine shop owner and had never before experienced financial difficulties. His later financial problems may have resulted from the types of inventions he worked on. Woods did not work on small consumer-type products but on large transportation systems. He often confronted powerful business people and politicians. People have speculated that the manager of the American Engineering Co. stole some of Woods's patents in 1892. As a result of defending his work to the public, Woods may have been charged with libel. He spent a short time in jail. The related legal defense all but forced him into bankruptcy, and he never fully recovered. He died of a stroke in 1910.

Woods never married, and little is known of his personal characteristics. The drawing on the first page of this profile is the one most commonly connected with him, though it is of uncertain origin. Like Edison, Woods devoted his technical career to invention. Also like Edison, his inventions came primarily in the field of electricity. In an October 1974 proclamation, Ohio governor John Gilligan recognized Woods as "the black Edison." It was a fitting tribute to an unsung genius of electricity.

References

The Real McCoy: African-American Invention and Innovation by Portia P. James, Smithsonian Institution Press, 1989.

Black Pioneers of Science and Invention by Louis Haber, Harcourt, Brace, & World, 1970.

Eight Black American Inventors by Robert Hayden, Addison-Wesley, 1972.

American Black Scientists and Inventors by Edward S. Jenkins and others, National Science Teachers Association, circa 1975.

Nikola TESLA

A 28-year-old Croatian immigrant reached New York City in the summer of 1884. He had just 4¢ in his pocket and many ideas in his head. The young man thought that he could easily use alternating current (ac), then in its infancy, to power electric motors. Nikola Tesla made the use of ac practical. We sometimes associate his name with the Tesla coil used in high-voltage, high-frequency electrical demonstrations. His most important contributions, however, involved the design of practical ac motors, alternators, and transmission equipment.

The fourth of five children, Tesla was born in a small village in what was then called Austria-Hungary. His birth was said to have come during a thunderstorm at the stroke of midnight. His father was a clergyman. His mother could neither read nor write, but she had an excellent memory and inventive skill in developing household devices. Tesla was close to his mother and credited her as the source of his creative ability. Static electricity and mechanical items fascinated the young man. He recalled stroking the back of a cat to produce a shower of crackling sparks. He also remembered building a tiny motor powered by June bugs that he attached to a cross on a spindle.

In the classroom, the tall, slender Tesla was an outstanding pupil, especially in mathematics. Without using pencil and paper, he often arrived at and blurted out the correct answers to complicated problems. After graduating from elementary school, Tesla moved in with an aunt and uncle in the larger town of Karlovac for his high school years. Many of his ancestors had had military or religious careers, and Tesla's father hoped that he would follow their lead.

Smithsonian Photo No. 80-16573

Nikola Tesla at age 29

Then, Tesla came down with a serious case of cholera shortly after graduating from high school. When it looked as if he would die, his father tried to rouse him by granting him permission to attend a technical school. Tesla recovered, spent a year rebuilding his strength, and entered the polytechnic institute in Graz, Austria.

The school had obtained a new direct current (dc) generator, or dynamo. The dynamo fascinated Tesla, and he spent many hours analyzing its operation. Noticing that its commutator caused a great deal of sparking, Tesla pondered the possibility of eliminating those electrical losses. His unusual

Born:
July 9, 1856, in Smiljan, Croatia

Died:
January 7, 1943, in New York, New York

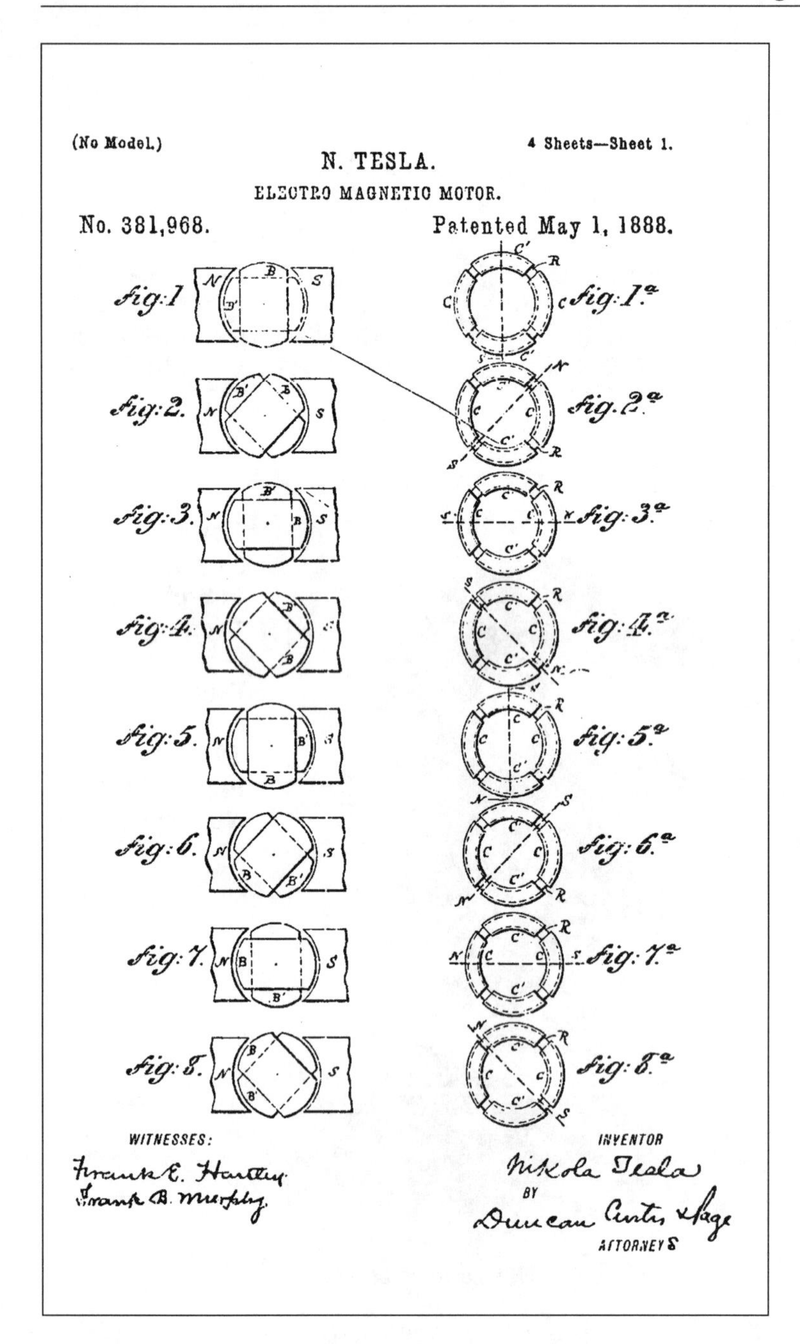

A Tesla ac motor patent showed the rotation of the armature.

intellectual powers allowed him to mentally design and test ac electrical devices. His instructors paid little attention to his ideas, which they thought were fanciful perpetual motion machines.

After receiving his degree, Tesla moved to Budapest to work as a draftsman. He again developed a grave illness and while recovering worked out designs for ac motors in his head. His few friends did not know how to respond to Tesla's seemingly strange comments which emerged from his mental testing of motors. After recovering from his illness, he redesigned dynamos and developed automatic regulators for an Edison plant in Paris. The job gave him contacts in America, and he decided to take advantage of the more challenging opportunities available abroad.

After arriving in New York, a letter of introduction got Tesla a position with Thomas Edison at his Pearl Street electrical generating plant in Manhattan. Tesla was a cultured man, while Edison was quite folksy. Yet, regardless of their differences, Tesla's work impressed Edison. Because Edison wanted to promote the use of dc power generation, he offered Tesla $18 per week plus a $50,000 bonus to significantly improve dynamo efficiency. Tesla took up the challenge, and he worked at it more than 16 hours a day, seven days a week, for several months. When he solved the problem at hand, Edison refused to pay the promised bonus. Edison said that Tesla obviously did not understand American humor. Angry and upset, Tesla quit on the spot. In 1912, when Tesla was offered the Nobel Prize in physics to be shared with Edison, he refused the honor because of his lingering anger toward Edison.

Tesla worked for several electric companies and experimented with ac motors and alternators in his spare time. He took out 12 patents in 1888, from his lifetime total of more than 100. He presented his ideas to the American Institute of Electrical Engineers and met with George Westinghouse. Unlike Edison, Westinghouse was committed to alternating current. However, he was having problems developing suitable motors and alternators. The devices covered by Tesla's patents filled the void and the two men made a deal. Tesla received $60,000 plus $2.50 for every horsepower of motor or generating capacity that Westinghouse sold. Tesla used the money to open his own private laboratory. Around that time, he also became an American citizen.

Westinghouse underbid Edison's company to win the contract to illuminate the 1893 world's fair in Chicago. This was the first fair to use electric lighting. After assembling 24 500-horsepower alternators, the smartly dressed Tesla gave demonstrations at the Westinghouse exhibit. He showed off an electric clock, fluorescent tubes, and huge noisy sparks from electric coils. His motors originally operated from an ac frequency of 133 cycles per second.

Finding that to be too high, he experimented until he came up with 60 cycles per second. That frequency was low enough to operate motors and high enough to eliminate lamp flickering.

The same year, Westinghouse won the contract to build three alternators at Niagara Falls. They would make up the world's first large hydroelectric generating plant. Three huge turbines were connected to 5,000 horsepower two-phase alternators designed by Tesla. The system went on-line in 1895, and a 22-mile ac transmission line soon carried power to Buffalo, New York. The plant was completely operational in 1902 with 21 alternators. By itself, it produced more electrical power than that generated in 31 states. Tesla's patents formed the basis of the entire power generation and transmission network.

In his ac-dc battle with Edison, Westinghouse spent large sums of money, bringing his company close to bankruptcy. He asked Tesla to modify their financial arrangement. Tesla agreed to sell his patent rights for $200,000. He was then at the height of his fame and felt he would get more money from other sources in the future. This was not a wise decision, but Tesla did not trust financial advisors. Lowering the royalty payments he received would have easily saved Westinghouse and still made Tesla a millionaire in a few years.

Tesla's pet project was the atmospheric transmission of electrical power. He worked on techniques to transmit power through the air for motors and lights. As part of that effort, he first worked on wireless communication. In 1897, he transmitted a wireless message over a distance of 25 miles. He patented and built a successful model of a radio-controlled ship in 1898. He was constructing a 187-foot tower on Long Island in 1901 just as Guglielmo Marconi sent the first wireless signal across the Atlantic Ocean. Tesla calmly commented, "Let him continue. He is using 17 of my patents." The U.S. Supreme Court agreed. In 1944, the justices voided Marconi's wireless patent in favor of a 1900 Tesla patent.

Tesla was clearly an underappreciated technical genius who had extraordinary powers of analysis and invention. But he was also an arrogant man who had little tact, had no friends, never married, and was almost uniformly disliked. Late in life, he became enamored with pigeons and spent hours a day feeding them. He often brought them into his New York apartment, where he died alone in 1943.

References

"Seeking Redress for Nikola Tesla" by Eliot Marshall, *Science*, 30 October 1981.

"Nikola Tesla" by Kenneth M. Swezey, *Science*, 16 May 1958.

"The Work of the World" by Curt Wohleber, in *American Heritage of Invention and Technology*, Winter 1992.

Dictionary of American Biography, Charles Scribner's Sons Publishers, 1932; with supplemental updates.

National Cyclopedia of American Biography, James T. White and Co. Publishers, 1891; with supplemental updates.

McGraw-Hill Encyclopedia of Biography, McGraw-Hill Publishers, 1973.

Asimov's Biographical Encyclopedia of Science and Technology, by Isaac Asimov, Doubleday & Co. Publishers, 1964.

Heinrich HERTZ

Born: February 22, 1857, in Hamburg, Germany

Died: January 1, 1894, in Bonn, Germany

In the 19th century, technically minded young people had a challenging career decision to make. On one side were the exciting and dynamic technologies such as steam locomotives and machine tools. On the other side were the emerging sciences like electricity and thermodynamics. Both sides had their champions, but that was little comfort for someone who had to choose a career. The situation has not changed much over the years. Today's young people have similarly challenging choices to make.

Some, like American scientist Josiah Willard Gibbs (1839-1903), avoided making a decision until they were almost 30. Others, like German communications pioneer Heinrich Hertz, tried their hand at both technology and science. Unsure of his future, Hertz started his college work in engineering but also took advanced science and mathematics courses. Technology and engineering were essentially identical career paths during the 19th century. The modern world is a better place because Hertz chose to study science. His discoveries were the first to blend electricity into electronics. Modern wireless electronic communication is used daily by almost everyone. It includes not only radio and television, but also telephones, fax machines, and the Internet. All those communication methods can be traced to an 1886 experiment Hertz conducted in southwestern Germany.

Hertz was the oldest of five children and had a comfortable childhood. His father was a lawyer and politician. His mother came from a family of physicians and religious leaders. His education started in a private school when he was six. The school's headmaster was quite demanding and Hertz's mother carefully watched over his homework. She wanted him to be first in the class, and usually he achieved that goal. Hertz's interest in technology developed after he received a workbench and hand tools when he was 12. His parents later purchased a lathe for him. Science had been his grandfather's hobby and Hertz received his grandfather's science equipment at an early age.

The Science Museum/Science & Society Picture Library

Graduating from high school in 1875, Hertz was unsure about his future. He worked for a construction organization in Frankfurt for about a year. He then briefly attended a technical school, Dresden Polytechnic. His year of military service was spent with a railway regiment in Berlin. Hertz moved to Munich in 1877 with the intention of continuing his engineering education at the Technische Hochschule, or Technical University. While preparing for his entrance examinations, he studied mathematics and science. He enjoyed those subjects so much that he decided to study physics at the University of Munich.

Hertz's professors were excellent teachers working in the new field of electricity. Hertz was an equally excellent student. He studied advanced mathematics during his first semester and did laboratory work during the second semester. German universities freely exchanged students and after a year of study Hertz was expected to move to another school. People called this arrangement *student migration*. Hertz chose the University of Berlin primarily because Hermann von Helmholtz (1821-1894) taught there. Helmholtz was Germany's leading scientist, with international recognition based on his work in the theory of electricity and magnetism. The Helmholtz coil is named after him. It develops very uniform magnetic fields. A person with several scientific interests, Helmholtz also invented the ophthalmoscope to study the human eye. He drew Hertz into his circle of students and the two developed a strong mutual respect.

To justify Helmhotz's acceptance, Hertz wanted to show his ability in a tangible way. His first technical triumph was winning a medal and a cash award offered by a professional society for the solution to a particularly difficult experimental problem. It dealt with the study of alternating currents. Helmholtz had proposed the problem and was interested in its solution. He provided Hertz with office space and regularly checked his progress. After months of effort, Hertz solved the problem, which was the basis for his first professional article. The solution guided him toward the discovery of radio waves.

Hertz enjoyed his work with electricity and wrote to his parents, "I am now thoroughly happy and could not wish things better." After receiving his doctorate degree in 1880, he stayed in Berlin to do additional work in electricity. He was Helmholtz's salaried assistant at the Berlin Physical Institute, a research facility. Hertz generated 15 publications and was developing a noteworthy reputation. His technical skills touched on both the theoretical and experimental aspects of electricity. Anxious to establish a career on his own, he took a teaching position at the University of Kiel in 1883.

Hertz was an excellent teacher and had more students than he expected. But his main interest was in conducting electrical research and Kiel had limited laboratory facilities. Karlsruhe Technical University had an opening in 1885. The school had a large laboratory and excellent test equipment. Hertz accepted the position when it was offered. At Karlsruhe, he conducted his landmark experiments in radio transmission.

While becoming familiar with the facilities, he met Elisabeth Doll, the daughter of a colleague who taught surveying. They married in 1886 and eventually had two daughters. The same year, Hertz conducted the experiment that would make him world famous. He had noticed that Leyden jars, an early form of capacitor, discharged with a spark across an air gap. A smaller spark would occasionally accompany the main spark across the terminals of a nearby Leyden jar. Hertz wondered if the main spark caused the other. After countless preliminary experiments, he constructed a high-voltage transmitter circuit that included two metal spheres separated by 1 to 2 cm. They were positioned at the focal point of a metal parabolic mirror. Electrical power came from a bank of wet-cell batteries and a can-shaped induction coil about 20 cm in diameter and 52 cm high. Relay contacts provided the make-and-break cycle to generate high voltage from the coil. The rapidly changing

Some of Hertz's original test equipment is displayed at Munich's Deutsches Museum. The frame-like devices are radio wave polarizers or antennas. His most successful early transmitter included the two wires with different diameter spheres at each end, shown at the bottom of the display case. Hertz called the device a linear oscillator.

A replica of Hertz's 1886 landmark radio transmission experiment is displayed at Munich's Deutsches Museum. An induction coil (far left) produces periodic sparking at contacts in the parabolic metal reflector. Electromagnetic waves are created and received by a similar reflector at the right. The waves pass through an eight-sided polarizer with gratings of wire. Hertz used the device to demonstrate the similarities between light and radio waves. The oscilloscope shows signal patterns to museum visitors.

current produced sparks that jumped the air gap between the metal spheres. The sparking created electromagnetic, or *radio,* waves. Hertz's receiver was a 70 cm diameter loop of wire with smaller metal spheres at the ends, separated by about 3 mm.

The equipment was crude, but Hertz obtained valuable data from each experiment he conducted. He placed the receiver at different parts of a darkened room, typically 20 meters from the transmitter. Large high-voltage sparking at the transmitter induced weak sparks to jump the air gap at the receiver. Hertz was the first person to send and receive energy through space. Nothing like this had ever been done before.

Hertz noticed a similarity between his electrical signals and light. He conducted a series of complex experiments in 1888 that showed that electrical signals were energy waves, just like light. Hertz reflected and refracted them. He focused them with different devices and calculated wave lengths. His equipment typically generated wave lengths of about 30 cm at a frequency of 100 MHz. He developed efficient antennas and discovered that short waves are better for information transmission than longer ones.

Hertz established the foundation for modern electronic communication. But he had not forgotten engineering and technology. He had a paper published in an engineering journal on a new method for measuring the hardness of metals. Hertz also developed an improved ammeter and a hygrometer for measuring humidity.

Hertz was an excellent teacher with a pleasant personality. He had strong language skills and spoke fluent English, French, and Italian, in addition to his native German. He also had a working knowledge of Arabic. His work resulted in the first long-distance transmission of a wireless communication. The Italian Guglielmo Marconi (1874-1937) sent a Morse code signal across the English Channel in 1899, a distance of about 30 miles. But Hertz did not live to see it happen. He contracted an infection from a decayed tooth and died in Bonn at the age of 37.

Some early technologists were brash and obvious. Examples are the American Isaac Singer (1811-1875) and the Briton Richard Arkwright (1732-1792). But they were more commonly quiet, low-key persons, like Hertz. In 1890, the British Royal Society awarded Hertz its coveted Rumford Medal, a precursor to the Nobel Prize. Hertz told no one when he traveled to London to accept it at ceremonies with the world's greatest scientists or after returning to Germany. He was modest to the point of never mentioning his discoveries, even when discussing their basic theory with students or colleagues. To honor Hertz's contributions, the international General Conference on Weights and Measures adopted the unit of *hertz* in 1960 as one cycle per second.

References

Pioneers of Electrical Communication by Rollo Appleyard, Books for Libraries Press, 1968.

The Communications Miracle by John Bray, Plenum Press, 1995.

Famous Names in Engineering by James Carvill, Butterworths Publishers, 1981.

Radio's 100 Men of Science by Orrin E. Dunlap, Jr., Books for Libraries Press, 1944.

Michael PUPIN

Born:
October 4, 1858,
in Idvor, Austria-Hungary

Died:
March 12, 1935,
in New York, New York

The U.S. Patent Office noted a significant increase in patent applications during 1924. Officials there credited the increase to the publication of Michael Pupin's autobiography, *From Immigrant to Inventor*, the previous year. A true rags-to-riches story about a poor Serbian-born genius in the field of electricity, it apparently encouraged many would-be inventors to make patent applications. The book was even more remarkable because Pupin won the 1924 Pulitzer Prize in biography for it. It was the only award of its kind ever received by a person whose career was grounded in technology.

Pupin was born to an illiterate but intelligent farm couple. His father served several terms as mayor of his small village and his mother strongly encouraged him in his studies. Pupin received his elementary education in Idvor, but he went to high school in a larger town nearby. His teacher introduced the young Pupin to electricity by conducting small demonstrations and explaining Benjamin Franklin's kite experiment. During summers at home, Pupin helped herdsmen at night with their oxen. They communicated with each other by using a long knife inserted deeply into the soil. Tapping on the knife's wooden handle sent vibrations through the ground. The technique gave Pupin his introduction to sound transmission.

While in his early teen years, Pupin made himself politically conspicuous by speaking out in favor of a new and rising feeling of nationalism. This activity brought him difficulties with some of the people around him. An earlier chance encounter with a traveling American couple had introduced him to some of the values considered im-

Library of Congress

portant by many in the United States. Pupin concluded that Americans must hold loftier ideals than those he saw around him. In 1874, following his father's death, the 15-year-old Pupin decided that his future lay across the Atlantic Ocean.

Pupin sold his clothes and books to buy a steamship ticket. He boarded in Hamburg and arrived in New York City 14 days later. He landed with 5¢ in his pocket and the clothes on his back. He could not speak English. He quickly found a job with a Delaware farmer who happened to be in New York. Pupin worked on the farm for a few years and started taking English lessons

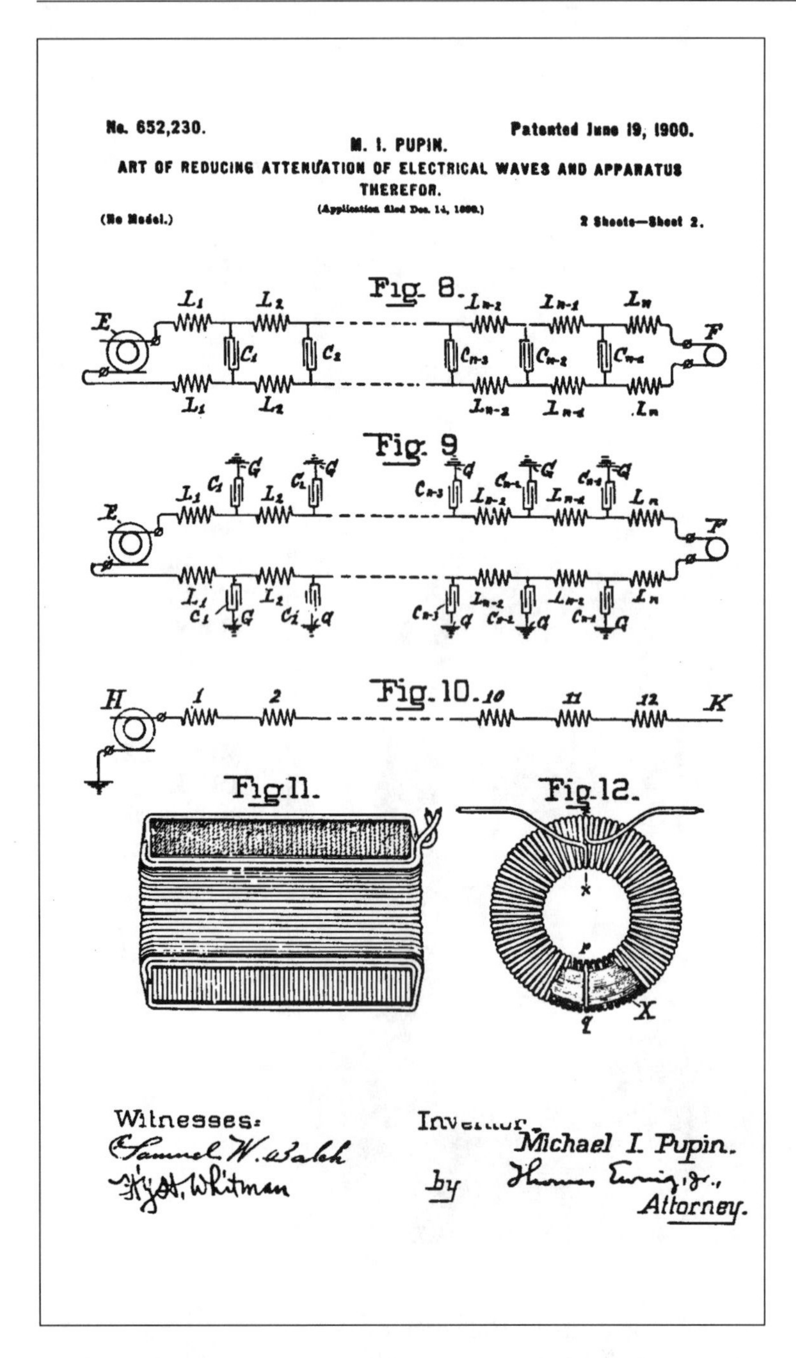

A patent drawing related to Pupin coils

from the farm owner's daughter. He later moved to a cracker factory on Cortland Street in New York, while acquainting himself with American customs and traditions. He roomed above the cracker factory with a German student named Bilharz who admired Pupin's desire to learn. Bilharz taught Pupin to read and write English, and later to read Greek and Latin. Pupin particularly enjoyed books on advanced mathematics and read each one several times.

Through dogged perseverance, Pupin prepared himself to apply for admission to Columbia University in 1879. He passed his entrance examination with honors and received a scholarship. Pupin involved himself so much with life in his new country that his classmates voted him class president his junior year. The week that he graduated from college in 1883, he became a naturalized citizen of the U.S. He then studied electricity further in England and Germany. Pupin returned to Columbia to become an instructor in the university's newly established electrical engineering department. He and Francis Crocker comprised the department, which had almost no equipment and no budget. The two men raised funds by giving lectures to business people and lawyers who had an interest in electrical industries. Pupin remained at Columbia for the remainder of his 39-year professional life.

The idea for Pupin's best-known invention came to him as he thought about a string with a weight on it. He knew that a disturbance set up in a loaded string does not die out as quickly as a disturbance in one that is not loaded. In his autobiography, Pupin said that what he learned with the herdsmen in Idvor about sound conducted through the ground helped him understand the weighted-string analogy. In the transmission of sound over telephone wires, particles are also set in motion, just like the string. Pupin thought that an electrically loaded transmission line would be more efficient than an unloaded line. Using that theory, he tackled the problem of telephone and telegraph transmissions that weakened and distorted as they traveled down a wire. Pupin did much theoretical work and experimentation to solve the problem. He determined that inductance coils, properly sized and strategically located along the phone lines, would greatly reduce the problem. The coils, spaced two to five miles apart and resembling wire-wrapped doughnuts, were called *pupinized lines* in Europe and *Pupin coils* in America. Pupin's discovery allowed people to hold intelligible conversations over distances as great as 1,000 miles. The Bell Telephone Co. bought his patent in 1901 and used it to make long-distance telephoning practical.

Pupin also did a great deal of work with medical X rays and made the first X-ray exposure in America in early 1896. It was of the hand of a young attorney who had been wounded by a shotgun blast. The X ray showed about two dozen shotgun pellets. By placing a fluorescent screen on top of the photographic plate, Pupin reduced long exposure times to just a few seconds. The fluorescent screen came from Pupin's close friend Thomas Edison.

Another of Pupin's discoveries involved the tuning of oscillating currents for radio reception. Every radio uses such a circuit to allow selecting the signal from only one station while excluding all others. Pupin was the first person to use the term "electrical tuning."

While working on his inventions, Pupin also taught electrical engineering courses. One of his students was Edwin Armstrong, who invented the frequency modulation (FM) system for radio transmission in 1939. He and Pupin jointly received a patent for a high-frequency vacuum-tube oscillator. Two of Pupin's students won Nobel Prizes. Robert Millikan won the award for physics in 1923 and Irving Langmuir won for chemistry in 1932.

Pupin married Sarah Jackson in 1888. Eight years later, he contracted tuberculosis. While caring for him, his wife also contracted the disease and died. A heart-broken Pupin moved to a farm in Connecticut to complete his recuperation. A friend presented him with a pair of horses on the condition that Pupin train them. Working with the horses brought Pupin out of his depression, and he returned to work several months later. Pupin bought the Connecticut farm and built a large home resembling a Serbian medieval landlord's house. After his death, it became the home of his only child, his daughter Varvara.

Much like Albert Einstein, Pupin worked in an area of technology that the average person could not understand. Still, in their times, both Einstein and Pupin were widely considered to be great men who deserved respect and recognition. Pupin received 34 patents, and he wrote 68 technical articles and 4 books. He received 5 medals and 18 honorary college degrees, 2 from European colleges.

Pupin was a large and generally healthy man who enjoyed hard work and athletic activity. He was an eloquent public speaker, with an upbeat personality and much social grace. Although he was a naturalized American citizen, Pupin never forgot his heritage. He organized an immigrant society and helped provide temporary housing for others from his homeland. He also helped to publish a newspaper of interest to immigrants. During World War I, Pupin represented the Serbian branch of the Yugoslavian government. He died on the 61st anniversary of the day his ship had left Hamburg. The Pupin Physics Laboratory at Columbia has a bronze plaque that honors him as "a true guide to the perplexed."

References

From Immigrant to Inventor by Michael Pupin, Charles Scribner's Sons, 1923.

Smithsonian Book of Invention, Smithsonian Institution, 1978.

Dictionary of American Biography, Charles Scribner's Sons Publishers, 1932; with supplemental updates.

National Cyclopedia of American Biography, James T. White and Co. Publishers, 1891; with supplemental updates.

McGraw-Hill Encyclopedia of Biography, McGraw-Hill Publishers, 1973.

Asimov's Biographical Encyclopedia of Science and Technology, by Isaac Asimov, Doubleday & Co. Publishers, 1964.

Herman HOLLERITH

Born:
February 29, 1860, in Buffalo, New York

Died:
November 17, 1929, in Washington, D.C.

The expanded use of computers in education, business, and industry might lead some to think that automated information retrieval is a modern innovation. It's not. The first system to store and analyze large amounts of data was invented in 1887. Herman Hollerith developed tabulating machines that could read and interpret information stored as holes in punched cards. Previously, French textile weavers had used punched paper and thin wood to automate the weaving of intricate fabric designs. Mathematicians had tried to make calculating machines using the same principle. However, it was Hollerith who first produced a practical information retrieval system that was the predecessor of electronic computers.

Hollerith was the seventh child of German immigrants. Raised for the most part by his widowed mother, he moved with his family to New York City at age 9. Hollerith's mother supported herself and her children by making women's hats. Hollerith graduated from Columbia University with a degree in mining engineering at the age of only 19, and his first job involved assisting one of his former teachers. William Towbridge had the responsibility of tabulating data for the 1880 United States' census. He assigned Hollerith to compile data on the statistics of manufacturers. The job had little to do with mining, but it paid $600 a year.

The project led Hollerith to Washington, D.C., where, on one occasion, he had dinner with a young lady and her family. Her father, John Billings, was a specialist at the Census Office. He discussed with Hollerith the desirability of having a machine that would do the routine work of tabulating

Smithsonian Institution Photo No. 060700

populations and other statistics. Billings speculated that the time required to conduct a complete census analysis by hand could take over 10 years. Counting by making marks with a pen worked fine for the first census of 1790, when the U.S. had only four million residents and census takers asked only four questions. Ninety years later, the U.S. population had multiplied by 12 times and citizens had to answer dozens of questions on the census. The 1880 census would take seven years to complete, and a more efficient tabulation method was an absolute necessity. His chance encounter with Bill-

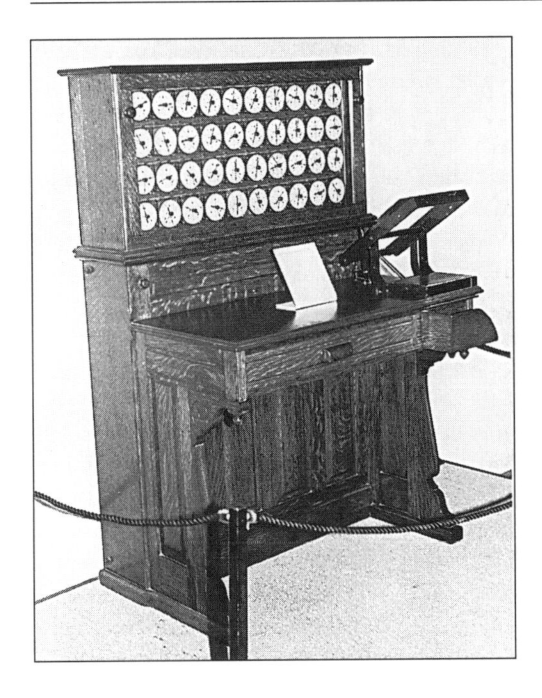

Full view of Hollerith's tabulator, in a museum display. The hand-operated card punch is on the desk at the right.

ings started the 20-year-old Hollerith thinking about a tabulating machine.

Finishing his duties with the 1880 census, Hollerith accepted a teaching job at the Massachusetts Institute of Technology in 1882. It gave him more time for experimentation, as well as the opportunity to talk with others about the field of applied electricity. His first serious attempt at creating a tabulating machine involved the use of a strip of paper with holes punched in it. Hollerith ran the paper over a metal drum and made contact through the holes to operate a counter. He liked the design because it was an automatic feed system. Hollerith patented his design, but never went into production with it because it proved difficult to make changes or locate data in long tapes.

After finding that he disliked teaching, Hollerith moved to St. Louis, where he worked for a short time on electrically controlled air brakes for trains. While traveling for work, he often noticed railroad conductors punching several holes in tickets. The holes were used for identification as a hedge against theft by vagrants who might try to resell the tickets. Hollerith wrote: "The conductor . . . punched out a description of the individual, as [having] light hair, dark eyes, large nose." These "punch photographs" gave Hollerith the idea he needed.

Hollerith moved to the Patent Office in 1884 and spent his spare time working on his tabulating machine. He worked for more than three years on different techniques, finding the use of punched cards the most versatile. A pin could be slipped through the punched hole into a tiny cup of mercury to complete an electrical circuit. Counting was automatic and registered on circular dials. When Hollerith was ready to build a prototype, he asked his brothers and sisters for financial help. None believed in his invention, and he received no assistance. After that rebuff, he had little contact with his siblings. He obtained the needed money in small amounts from many other investors.

Hollerith's tabulation method had to compete with two others for the 1890 census contract. The others would have used the conventional method, in which clerks worked with tally sheets, colored paper, and colored pens. Hollerith's technique was so unusual that officials decided to test it with Baltimore's 1887 death records. The cards were punched with a train conductor's ticket punch. Hollerith recorded age, sex, and cause of death in grid squares printed on thousands of cards. His battery-powered tabulator completed the analysis in a few days instead of the several weeks required by hand tabulation. Amazed at the system's speed, officials asked Hollerith to repeat the process for New Jersey and New York City. He did so, and won the contract for the 1890 census.

Hollerith ordered cards the same size as dollar bills, which allowed them to fit into

Hollerith's keypunch machine rests on the working surface of his battery powered tabulator. The dials in the background recorded totals as the punched cards were fed into the tabulator.

This page of Hollerith's patent shows a process for punching holes in paper, a stage of processing data.

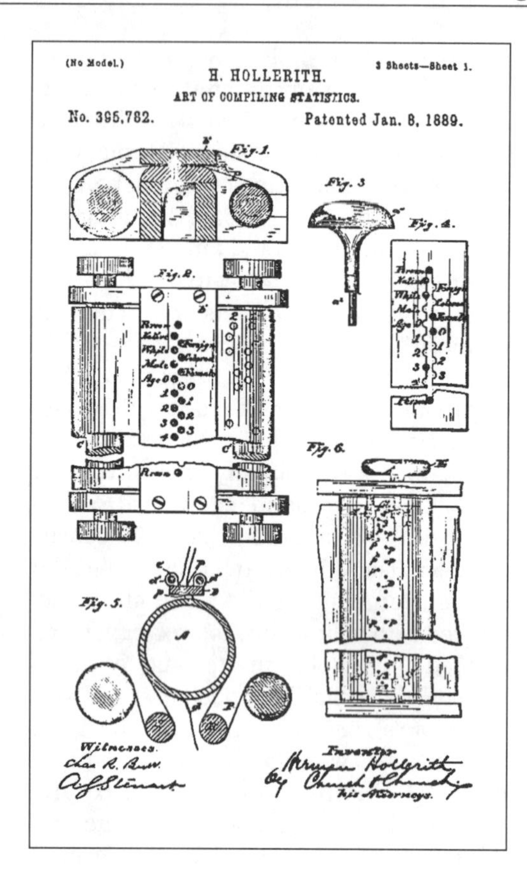

file drawers used by banks. Instead of using a single-hole hand punch, Hollerith invented a keypunch machine. The photograph on the preceding page shows one on the working surface of a tabulator that could handle up to 40 pieces of information per card. Hollerith started with an initial 30 machines that the Western Electric Co. built for $1,000 each. Up to 100 women during the day and 100 men at night punched information into the cards. Workers fed cards individually into the machines, and the dials behind the keypunch registered totals. It took Hollerith only six weeks to arrive at a figure of 62.5 million as a rough count of the U.S. population. When the final tabulations came in just two years later, the actual total was 63,056,000. The entire process had consumed only one third of the time it was estimated to take.

The following year, Canada, Norway, and Austria used Hollerith's equipment for their censuses. Russia's first-ever census was also tallied on Hollerith equipment. Hollerith attended the International Statistical Institute in 1895 in Bern, Switzerland. Over the next several years, he adapted his machines to handle statistical data. He found an expanded market and was soon making equipment for business, industry, and education.

Hollerith established the Tabulating Machine Co. to manufacture machines and sell punch cards. Not forgetting how he got started, he offered Billings a share in the business. Following the lead of George Pullman, who rented—rather than sold—his fancy railroad cars, Hollerith rented his machines. Within a few years, hundreds of companies were renting Hollerith punched-card tabulators. He sold the company in 1911 for an estimated $2 million. In 1924, its name changed to the International Business Machines Corporation.

In his lifetime, Hollerith was granted 38 patents and became a wealthy man. He took pleasure in the gold medal his tabulator received from the French government in 1889. He also received the prestigious Elliott Cresson medal from Philadelphia's Franklin Institute in 1890. He lived with his wife and six children in expensive houses in the Georgetown section of Washington, D.C., about a mile from the White House. He owned four automobiles, a yacht, and country acreage. Because he was born on February 29 of a leap year, he had only 17 birthdays in the course of his lifetime.

References

Those Inventive Americans, National Geographic Society, 1971.

Machines—LIFE Science Library by Robert O'Brien, Time Inc., 1964.

Dictionary of American Biography, Charles Scribner's Sons Publishers, 1932; with supplemental updates.

Charles STEINMETZ

Electricity was the most sophisticated emerging technology during the 19th century. The use of electricity for mechanical power was a particularly challenging field. Direct current (dc) for electric motors was commonly used, but alternating current (ac) was less well understood. An immigrant who gained worldwide recognition for his work in electricity stood just 4' tall, was hunchbacked, and had misshapen legs. Yet, the disadvantages of poverty and a crippling deformity did not keep him from being recognized as a technical genius. More than any other single person, German-born Charles Proteus Steinmetz was responsible for establishing the United States' early leadership in the field of electricity.

Steinmetz was born in Breslau, Germany, a city that is now called Wroclaw in present-day Poland. In spite of his deformities, he had remarkably good health and he had a relatively normal childhood. His father headed the printing department in the local railroad office. Steinmetz's mother died during his first year, and his grandmother, who spoiled him badly, raised him.

It was obvious that Steinmetz had a keen technical mind. After he graduated from high school, his father sent him to the University of Breslau, instead of considering an apprenticeship. During his years at the university, Steinmetz never missed a class. He took lengthy class notes and often did independent investigations at home. Although his father encouraged him to attend college, he could not offer much financial help. Steinmetz made money by tutoring and doing odd jobs. It took him six years to complete his studies. He always selected the most difficult technical subjects. During his last year at the university, he specialized in six major subjects: electrical engineering, mathematics, physics, chemistry, astronomy, and medicine.

General Electric Company

Charles Steinmetz (birth name: Karl August Rudolph Steinmetz) working at his desk in 1923

Born:
April 9, 1865, in Breslau, Germany

Died:
October 26, 1923, in Schenectady, New York

In Steinmetz's time, students at European universities commonly joined one or more outside societies. Steinmetz enjoyed the personal contacts. Because he thought that there was much social injustice in the world, he affiliated himself with the Social Revolution Party in the late 1880s. As student members of the group grew more politically active, they found themselves under suspicion. When the authorities arrested about 40 of Steinmetz's friends, he knew he would soon

Steinmetz's summer cabin is on display at Greenfield Village in Dearborn, Michigan. This is the first section he had built in the 1890s and it was originally on stilts. Three other rooms were later added to it.

join them. Shortly after he completed his university studies, he hurriedly left home on the morning he was to be arrested. He was 24 years old, and he never saw his family again.

Secretly arriving in Zurich, Switzerland, Steinmetz became friends with Oscar Asmussen. Asmussen was wealthy, and he offered to pay Steinmetz's boat fare to America. When the two men arrived in New York, immigration officials at first wanted to return the penniless Steinmetz to Europe. He spoke no English and thus could not defend himself verbally. However, when Asmussen showed a large amount of money, telling the officials that it belonged to both of them, they allowed Steinmetz to enter the U.S.

Steinmetz found work in Yonkers, New York, with the electrical inventor Rudolph Eickemeyer. Eickemeyer was developing dc electrical motors and electrical machinery. He found Steinmetz's keen technical mind to be of great help and set him up with a laboratory. Steinmetz began working on the problem of hysteresis, the loss of electrical motor efficiency caused by alternating magnetism. The laws of power loss were unknown, and many people did not believe hysteresis existed. No one had been able to measure it. Using existing data, Steinmetz applied high-level mathematics to show that it did exist. He also demonstrated how to measure and reduce hysteresis. He presented his results to the American Institute of Electrical Engineers in 1882 and became an immediate sensation in technology. He showed people how to deal with the complicated effects that resulted from the use of ac. The methods Steinmetz used were so complete that a motor designer who understood them could hardly make a mistake.

Steinmetz had found a country that accepted him as he was, and he felt comfortable in America. Shortly after arriving, he had learned the language so well that he spoke with only a slight accent. He became an American citizen in 1894 and changed his first name to Charles. His middle name, Proteus, was his college nickname. It came from Greek mythology and referred to someone who could quickly change character.

The General Electric Co. (GE) was formed in 1892. It began through the consolidation of the Edison General Electric Co. in Schenectady, New York, and the Thomson-Houston Electric Co. in Lynn, Massachusetts. Its main offices were in Schenectady. Those in charge at GE wanted to obtain as much talent in the field of electricity as they could. They invited Steinmetz to join their staff, but he refused. He did not want to turn his back on the employer who gave him his start in America. In its most important action in the 19th-century, GE offered to buy Eickemeyer's company with the understanding that GE would also receive Steinmetz's services. Everyone agreed and Steinmetz soon moved to Schenectady, where he remained for the rest of his life.

After Steinmetz's arrival at the company, GE established research laboratories. They were the first company-sponsored industrial research facilities in the U.S. While still in his twenties, Steinmetz joined as a member of the calculating department. He soon became a consulting engineer. That position allowed him to come and go as he pleased, and he held it from then on. Although well liked and highly respected, he did work that almost no one could understand. Through the publication of 10 technical books, Steinmetz gradually made ac electricity less mysterious. He showed the profitability of applying advanced mathematical methods to practical electrical problems. Over his lifetime, he obtained 195 patents, and his methods are now universally used in ac equipment design and calculations.

Steinmetz conducted much of his theoretical work at a cabin he had built on a tributary of the Mohawk River. He would often work there for weeks at a time. He typically wore a bathing suit and a faded red sweater, which he did not change for even the most important visitors. After a day in his canoe, he once wrote, "It was a hot sunny day with almost no wind, and I sat in the sun and calculated instances of condenser discharge through an asymmetrical gas circuit." As with Isaac Newton and Albert Einstein, the public did not understand Steinmetz's work. However, everyone acknowledged and respected his technical contributions.

Although Steinmetz had completed the necessary academic work, the University of Breslau had never granted his degree. He felt pleased when Union College in Schenectady awarded him an honorary Ph.D. in 1903. From then on, Steinmetz used the title of "doctor." He also received many other forms of recognition. One of the more unusual was his election as head of the local board of education for two terms.

Steinmetz's favorite hobbies included photography, music, and bicycling. He also enjoyed botany, particularly work with desert plants, orchids, and ferns. He had a large greenhouse attached to his house. He rarely traveled, preferring to spend time near home. Steinmetz frequently smoked foul-smelling cigars and gave the immediate impression of being gruff and unapproachable. Yet, just the opposite was true.

Steinmetz loved children, though he had no family of his own. In 1905, he legally adopted the family of his young lab assistant, Joseph Hayden. Hayden, his wife, and their three children shared Steinmetz's large rambling home on Schenectady's Wendell Avenue. Steinmetz died of heart failure shortly after returning with the Haydens from a tour of the West Coast, his first long vacation trip.

References

Recollections of Steinmetz by Emil J. Remscheid, GE Co., 1977.

The Life of Charles Proteus Steinmetz by Johnathan Norton Leonard, Doubleday Publishers, 1929.

"The Mentor" by John Winthrop Hammond, *Charles Proteus Steinmetz*, May 1925 (Vol. 13, No. 4). (Supplied by GE, Schenectady, NY.)

Dictionary of American Biography, Charles Scribner's Sons Publishers, 1932; with supplemental updates.

National Cyclopedia of American Biography, James T. White and Co. Publishers, 1891; with supplemental updates.

McGraw-Hill Encyclopedia of Biography, McGraw-Hill Publishers, 1973.

Asimov's Biographical Encyclopedia of Science and Technology by Isaac Asimov, Doubleday & Co. Publishers, 1964.

Lee DE FOREST

Born: August 26, 1873, in Council Bluffs, Iowa

Died: June 30, 1961, in Hollywood, California

Early radio signals were transmitted from point to point and consisted only of dots and dashes. Average people had difficulty deciphering Morse code, and early radio was not a source of entertainment. It was Lee De Forest's inventive genius that transformed radio transmission to accommodate both voice and music. In 1907, De Forest invented the triode, the first vacuum tube that could faithfully reproduce audible sound. Most people think that Guglielmo Marconi invented the radio, but he did not. Marconi invented wireless telegraphy. De Forest invented radio. His audion plucked voice and music from the air and delivered it to the human ear.

When De Forest was six, his minister father became the first president of Talladega College in Alabama. Talladega was a newly founded college for children of former slaves. As a boy, De Forest grew up in the midst of racial tension that he could not understand. He had few friends, and his imagination led him to books on mechanics. The young De Forest constructed wooden models of such things as steam locomotives and blast furnaces. During part of the summer of 1893, he worked at the World's Columbian Exposition in Chicago. The time he spent at the exposition's Machinery Hall persuaded him to pursue an education in science and technology. His father had hoped he would become a minister.

Smithsonian Institution Photo No. 52213

Lee De Forest during a broadcast of the Metropolitan Opera about 1910

The following autumn, De Forest received a scholarship and enrolled in Yale University at New Haven, Connecticut. Although he started out in mechanical engineering, he took one of the first courses on the subject of electricity ever offered at a university in the United States. De Forest earned a doctorate degree, writing his dissertation on the reflection of radio waves. He worked at several different electrical-related jobs, but he did not like any of them. After that, Chicago's Armour Institute provided him with a small teaching salary and laboratory space for working on his experimental devices. Then, just three years out of college, De Forest persuaded several stock promoters to establish the American De Forest Wireless Telegraphy Co. He demonstrated that his equipment could transmit a dot-dash signal at least six miles. The U.S. government, which wanted to gain independence from foreign wireless companies, gave De Forest several contracts. Also, a fruit company decided to build a chain of his radio stations between Costa Rica and Panama. De Forest's company built 90 stations before he discovered that executives were mishandling company funds. He left the company and started another. De Forest repeated this pattern of establishing one company after another several times during his life.

In the early 1900s, wireless communication used Morse code, a series of dots and dashes generated by powerful sparks. The sparking required the use of six large, heavy electrical condensers. A typical one-ton condenser fit inside a 2' x 7'-long wooden box. Glass plates and kerosene filled each box. De Forest set out to simplify and improve the bulky system, which had the danger of producing shocks and fire.

De Forest worked on methods for improving vacuum-tube diodes. He set up an experimental station at the 1904 World's Fair in St. Louis. It included a 300'-tall antenna, the tallest structure at the fair. When he succeeded at sending wireless messages to Chicago, De Forest received the grand-prize medal for general excellence in wireless telegraphy. In 1906, he sent a 1,000-word message from Coney Island to Ireland—a distance of 3,400 miles. More than 570 of the words were received. It was the first such transmission since Marconi's transmission of three dots in 1901.

Like most electrical investigators, De Forest was searching for a more sensitive detector. He wanted to invent a device or circuit that would receive transmissions from long distances. After much experimental trial and error, De Forest placed a zigzag-shaped piece of nickel wire between the anode and cathode of a diode. He called the wire a *grid*. With proper circuitry, the three-element tube amplified Morse code far better than any diode could. De Forest called his tube the audion, or *triode*, and took out a patent in 1907. It was the prototype for billions of radio tubes that followed.

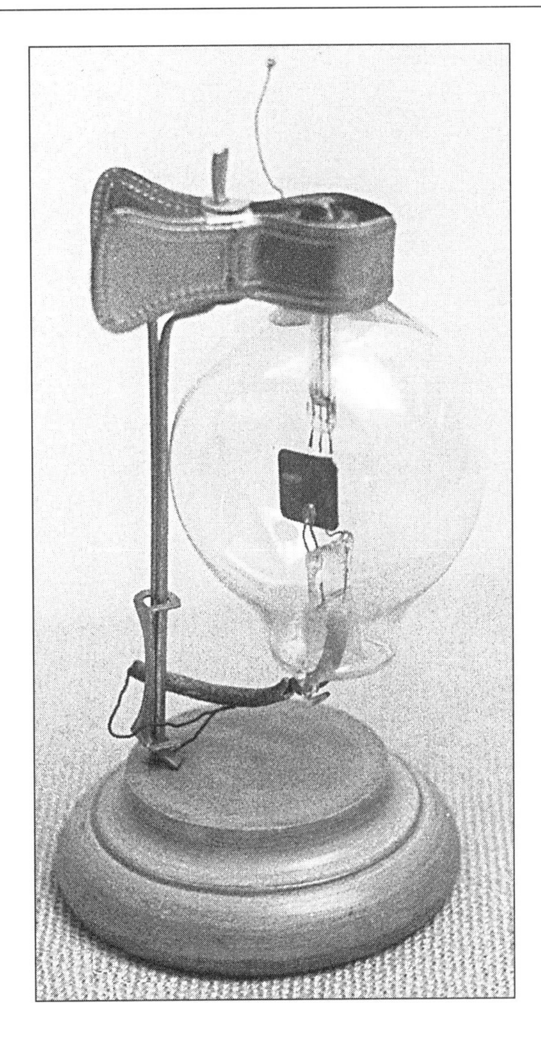

An early production model of De Forest's audion

De Forest did not appear to understand the capabilities of the triode. The theories he proposed were incorrect. Not until five years later did he realize the triode's most significant feature: It could amplify voice.

Using complicated electrical connections, De Forest produced a feedback circuit and generated musical tones. The tones could be varied to transmit and receive audible sound. Availability of De Forest's feedback circuit allowed the creation of the broadcast industry, and wireless communication was no longer limited to transmitting Morse code. De Forest also discovered that he could connect the output of one triode to the input of another. Through such a staging technique, several triodes could amplify weak signals.

Instead of point-to-point transmission, De Forest was more interested in *broadcasting* radio signals. (The term broadcasting comes from farming and refers to "spreading widely" in planting seed.) Thirteen years

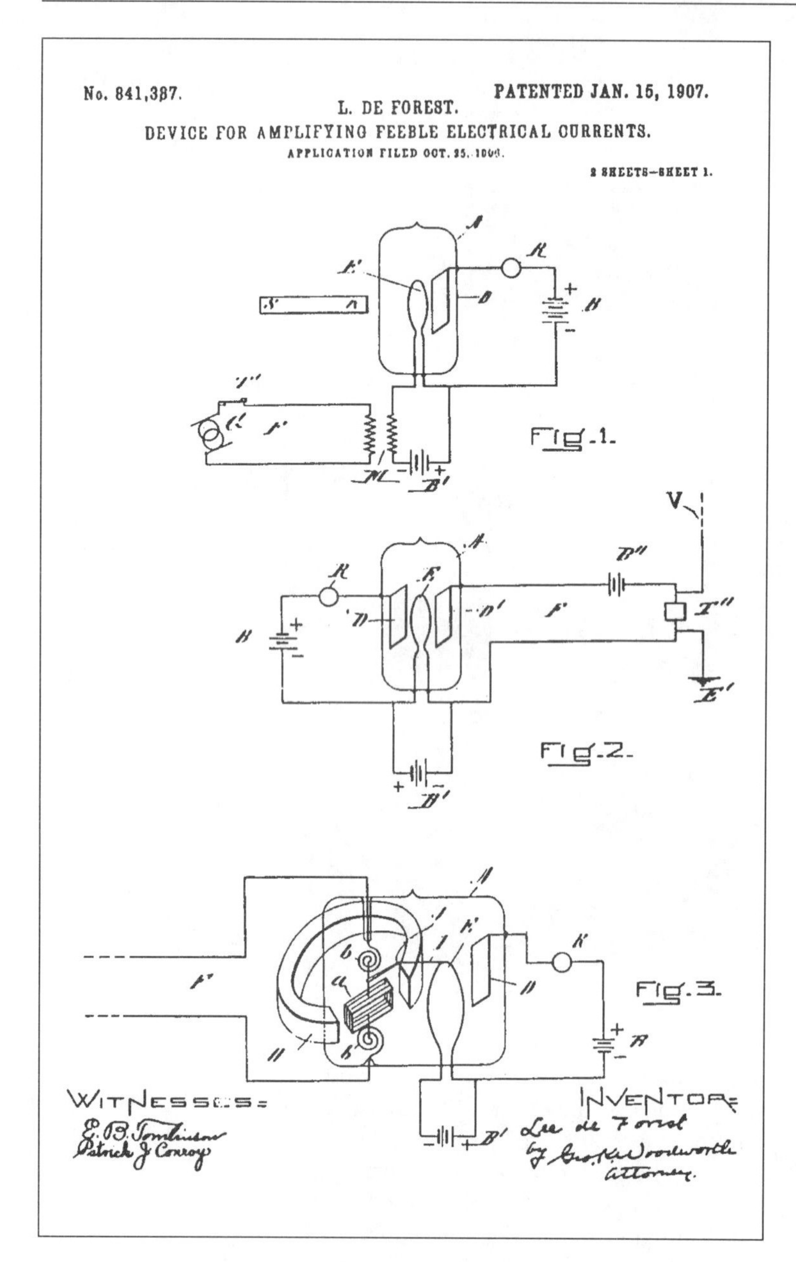

De Forest's audion spawned an increase in radio communication.

before the first radio station went on the air, De Forest arranged a broadcast of New York Metropolitan Opera singer Enrico Caruso. However, the early equipment was not up to the task, and the sound was almost inaudible. De Forest's concept was far ahead of its time.

De Forest worked during a period when most people were unsure of the existence of electrons. No one quite understood how the audion worked, and some people claimed they had previously discovered De Forest's feedback circuit. Those two factors resulted in one of the longest patent battles in history. The two main combatants were De Forest and Edwin Armstrong. In 14 years of court battles, De Forest won seven times and Armstrong won six times. It was not until 1934 that the U.S. Supreme Court finally decided in De Forest's favor. During the lengthy court proceedings, his audion made fortunes for others, helped create the Radio Corporation of America (RCA), and brought a new form of entertainment to the public. De Forest received little wealth or fame, and much unhappiness, as a result of his invention.

De Forest had pioneered sound transmission, but he ran into legal or financial problems at almost every stage of his work. Whenever he tried to manufacture or market his inventions, he encountered lengthy patent suits. While in his fifties, De Forest changed his strategy. He devoted himself to invention and then sold his rights to others. Only in his later years did he find contentment.

De Forest married four times. His fourth and happiest marriage, in 1930, was to silent-movie star Marie Mosquini. She shared his enjoyment of music, poetry, and camping. De Forest spent little time in the laboratory during his later years. He preferred to listen to music from the large high-fidelity phonograph he built for his own use. He received more than 300 patents over the course of his lifetime, with the last granted when he was 83. Many were commercially successful, but the triode had by far the greatest importance. It was one of this century's greatest inventions. Late in his life, De Forest called his audion "the granddaddy of all the vast progeny of electronic tubes that have ever come into existence."

References

Empire of the Air by Tom Lewis, Harper Collins Publishers, 1991.

Those Inventive Americans, National Geographic Society, 1971.

American Science and Invention by Mitchell Wilson, Bonanza Books, 1960.

Dictionary of American Biography, Charles Scribner's Sons Publishers, 1932; with supplemental updates.

National Cyclopedia of American Biography, James T. White and Co. Publishers, 1891; with supplemental updates.

Asimov's Biographical Encyclopedia of Science and Technology, by Isaac Asimov, Doubleday & Co. Publishers, 1964.

Guglielmo MARCONI

Born:
April 25, 1874, in Bologna, Italy

Died:
July 20, 1937, in Rome, Italy

Telegraphy was the only means of rapid long-distance communication in the 1800s. It went underwater in 1865 when a thin wire connected North America with Europe. That was the year Cyrus Field (1819-1892) completed the first permanent transatlantic telegraph cable. The project required seven attempts before it succeeded. So it is not surprising the telegraph company charged $5 to $10 a word to send messages to Europe. The surprise is that there were plenty of customers.

Werner Siemens (1816-1892) spent many years connecting London, England, to Calcutta, India, by telegraph. Work on the 6,000-mile distance was finished in 1870. Had wireless communication been available, Field and Siemens might not have tried such huge projects. The first person to send and receive a wireless communication signal was Guglielmo Marconi in Italy. His most high-profile technical triumph occurred in 1901, when he sent the first transatlantic wireless signal from England to Newfoundland.

Courtesy Deutsches Museum, München

Marconi was born into a wealthy family that enjoyed all the trappings that money could buy. The Marconis had a townhouse in Bologna and an estate outside the city named Villa Grifone. Marconi's father was a landowner. His mother was the daughter of a whiskey distiller in Dublin, Ireland. Between the ages of three and six, Marconi lived in England with his mother and older brother, Alfonso. When he returned to Italy, he could barely speak his native language. Because he spoke Italian poorly and with an English accent, dressed well, and did not enjoy sports, his classmates picked on him. His father decided to educate him with private tutors.

Marconi's mother was 17 years younger than his father and she enjoyed social activities. She traveled extensively with her sons, sometimes staying for long periods of time in various European cities. Marconi later became a world traveler. His mother had given him the social skills to move easily among different cultures.

The mother often took her sons to Livorno on the west coast of Italy. When Marconi was 13, he began attending the Technical Institute in that city. He learned that Heinrich Hertz (1857-1894) had recently sent electricity through the air. Marconi became intrigued by the possibility of wireless telegraphy. He had a weak educa-

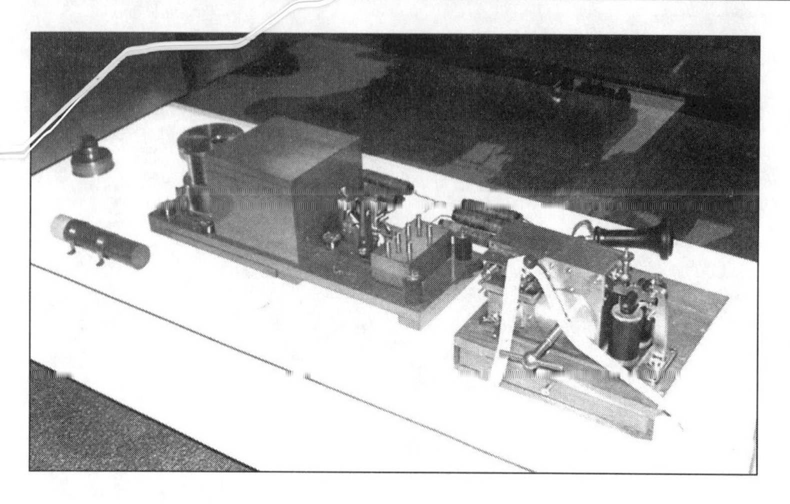

Replica of Marconi's 1901 transatlantic receiver. A kite-borne antenna brought the signal into the receiver and Marconi heard it through a telephone earpiece like the one at right center.

tional background and failed entrance examinations for the University of Bologna in 1894. His mother asked a neighbor and science professor, Augusto Righi, if her son could unofficially use the university's laboratory facilities. Righi agreed. Marconi's father disapproved and wanted his son to follow a career in the Italian Navy. He gave little encouragement to his son's interest in technical subjects.

Using university resources, Marconi set up a crude laboratory in the attic of the Villa Grifone. Righi cautioned Marconi that he did not have the educational background to succeed at work that had baffled scientists for years. But the young man was undaunted and soon had some success sending signals indoors. He showed his father that he could ring a bell at the opposite end of the house. A wireless signal tripped the bell's relay. His father was so impressed that he gave Marconi $1,000 for additional equipment.

One day in 1895, the Marconi brothers went outside with an oscillator, coherer, meters, switches, antennas, and other pieces of equipment. The oscillator produced sparks that Marconi hoped to receive at a distant point. The receiver included a coherer, a small four-inch glass tube filled with metal filings. The coherer conducted electricity only when it received an electromagnetic signal. The conductivity could be read by the needle of a galvanometer.

Marconi stayed with the transmitter. Alfonso went a mile and a half away with the receiver and a rifle. He would fire the rifle if the receiver picked up a signal. Marconi closed the switch. Alfonso received the signal and fired the rifle. That experiment is often cited as a first in wireless transmission.

The Italian Ministry of Posts and Telegraph showed no interest in funding more advanced experiments. Like most others, officials there felt wireless telegraphy would only be useful for ship-to-shore, or ship-to-ship communication. They suggested that Marconi try to find assistance in England, the world's most important seafaring nation. Marconi's mother arranged a meeting with William Preece, chief engineer of the Great Britain Post Office. Marconi was an almost unproven 21-year-old. Nonetheless, Preece was impressed and recommend that the British government support his work.

Using government and family money, Marconi established a research organization and surrounded himself with world-class technical experts. One was John Ambrose Fleming (1849-1945), inventor of the 1904 tube-type diode. Marconi's goal was to send and receive signals at greater distances. He first worked on a tuning circuit. One small region of the country might have many transmitters. If several signals were sent at the same time, the receivers could not separate the transmissions. The effect would be like a roomful of people all talking at once. Marconi patented his tuning circuit in England in 1900. Numbered 7,777, it allowed for transmitting radio waves at specific frequencies. The effect was like a person concentrating on a single speaker.

Marconi established the Wireless Telegraph and Signal Company to market his products. His first major sale was to the British government for use in the 1899-1902 Boer War. Partly for the publicity, Marconi used his equipment to transmit nine miles across the Bristol Channel in 1897. Interrupting high-voltage spark signals sent Morse

Replica of Marconi's 1901 transatlantic transmitter. Closing the key (right) energized the coil (center) and charged the two enclosed brass spheres (left). The spheres discharged through the antenna, sending an electromagnetic wave. The system was powered by a bank of batteries.

code. Marconi transmitted at about 15 words per minute. He then succeeded in transmitting 28 miles across the English Channel in 1899. But the greatest challenge involved sending electromagnetic waves across the Atlantic Ocean.

An unlimited capacity for work and complete faith in his abilities helped Marconi in his efforts. Assisted by Fleming, he set up a transmitter near Lizard Point, England, in the southwest corner of the country. His receiver was in St. John's, Newfoundland, about 2,100 miles away. The transmitter's antenna consisted of two 150-foot poles, placed about 170-feet apart, and strung with 55 copper wires. But the receiver's antenna was simpler: a piece of wire flown from a kite. Both the transmitter and receiver were far more complex than what Marconi had used in Villa Grifone just a few years earlier. The equipment and expenses brought the experiment's cost to $200,000.

Marconi's lifelong friend Luigi Solari operated the transmitter. Solari sent three dots at specific times, Morse code for the letter "S." In Newfoundland, Marconi listened through a telephone earpiece. At 12:30 P.M. on December 12, 1901, he barely heard three clicks. He passed the earpiece to his assistant George Kemp to verify the reception. When the news broke, Marconi's name was suddenly known throughout the world.

Marconi went on to achieve many other technical accomplishments. They were more and more detailed as the complexity of electronic communication became apparent. His business interests assumed international proportions. Marconi's United States branch became the Radio Corporation of America. From 1919, he lived on a large 220-foot-long yacht named *Elettra*. He bought it from the British government who had obtained it from Austria during World War I. Marconi modified the 730-ton yacht to carry tall wireless antennas and outfitted a complete shipboard laboratory. His work was acknowledged with the highest recognition. Marconi received the 1909 Nobel Prize in physics. He shared it with Karl Braun (1850-1918), German inventor of the cathode-ray oscilloscope.

Marconi's three-circuit multiple tuner. His installations from 1907 to about 1914 used this unit to couple the antenna to the receiver. Its three tuned circuits, loosely coupled together, gave good selectivity for a pre-electronic device. The three cylinders on top are condensers. The front dial at left varies antenna inductance and the unusual front dial with bars at right is for tuning.

Marconi married twice. First to Beatrice O'Brien from Ireland and then to Maria Cristina Bezzi-Scali from Italy. He had four children. He enjoyed fishing, horseback riding, and traveling. He was a careful dresser and almost all his photographs show his concern for clothing. An automobile accident took his right eye in 1912 and he had an artificial one fitted. It is all but impossible to detect in photographs of him. Marconi died of a heart attack at 63 while preparing for an evening appointment with Italy's leader Benito Mussolini (1883-1945).

As important as Marconi's contributions were, he did not invent radio. He invented point-to-point wireless telegraphy, not voice transmission. Electronic communication progressed rapidly during Marconi's lifetime. Only 27 years after he sent three barely audible clicks across the Atlantic Ocean, John Logie Baird (1888-1946) sent the first television signal from London to New York City.

References

Marconi by W. P. Jolly, Constable and Co. Publishers, 1972.

Marconi, The Man and His Wireless by Orrin E. Dunlap, Macmillan Publishers, 1937.

Great Lives from History edited by Frank N. Magill, Salem Press, 1987.

The Making of the Modern World edited by Neil Cossons, Science Museum Publications, 1992.

John BAIRD

Born:
August 13, 1888, in Helensburgh, Scotland

Died:
June 14, 1946, in Bexhill, England

Backpackers are adventuresome but generally stick to established trails. Every once in a while, they strike out on their own on an unmarked trail. Such treks are seldom troublesome because the backpackers carry maps based on information from people who preceded them. The field of technology has no maps and inventors must use their best judgment. Thomas Edison (1847-1931) spent a long time looking for a practical filament for his incandescent lamp. During that trek, he once famously said that he had discovered 2,000 materials that would not work.

Technologists began investigating the concept of television before electronics was even discovered. They first worked at sending a nonmoving image down a telephone wire. This involved breaking the image into small pieces. Current terminology would call that "digitizing the image." At the destination, the pieces would be reassembled, much as they are in a jigsaw puzzle. Fax machines operate that way. In an era before dependable electronics, one inventor developed a transmission and reception theory based on a spinning disk. It came to be called *mechanical television* and John Logie Baird was the first to make a practical system. Baird sent the first live television transmission in 1925. His equipment was the first used in broadcast service in 1929. The British Broadcasting Corporation (BBC) granted him the world's first television license. The BBC may have been influenced by Baird's accomplishment the previous year. In 1928, he sent the first transatlantic television image from England to New York. But his method followed an unmarked trail that had a dead end.

Rendering by Tim Harmon

Baird was born about 25 miles west of Glasgow in Scotland. His birth city was across the River Clyde from James Watt's birth city, Greenock. Baird was the youngest of four children. His father was a poorly paid minister and the family was not particularly well off. Baird was a sickly child and suffered from severe respiratory ailments all his life. He attended the local schools and read all the technical books and popular magazines he could find. While in his teens, he wired his house for electric lamps. The lamps were battery powered and recharged with a generator run by flowing water. As a youth, Baird was also interested

Baird's laboratory was located on an upper floor of this London building at 22 Firth Street. The first floor houses a restaurant named Bar Italia and a circular plaque opposite the clock commemorates the historical location. It states, "In 1926, in this house, John Logie Baird first demonstrated television."

in telephone operation and photography.

Baird started studying electricity at the Royal Technical College in Glasgow when he was 18. He heard about the early investigations of transmitting images by wire but did not show interest in the subject at that time. Baird was more concerned with gaining an education so he could make a living. He found employment in Glasgow at an automobile company and then at an electrical power company. Both jobs were hard on his health and he often missed work because of chronic illnesses. When World War I broke out in 1914, Baird was rejected for service. He made ends meet by selling socks and soap to department stores. He even moved to the warmer climate of Trinidad in the Caribbean Sea to make jams and jellies for sale. None of these ventures succeeded. Bad luck dogged Baird through his early years. At 34, he was sick, jobless, and had only £200 to his name. He considered himself a failure.

Then, Baird moved in with a childhood friend, Guy Robertson, in Hastings, England, about 60 miles south of London. He took long walks and his health began to improve. Baird felt there was only one way to overcome his failures: he must invent something. He remembered his youthful enthusiasm for electricity and the discussions of television at school. There had been many improvements in electronics over the past few years and Baird began to read technical publications again. He decided to try to make a practical television system, a goal that had eluded everyone else.

In the early 1920s, most investigators used a form of the Nipkow disk as an experimental transmitter. The Nipkow disk was a revolving metal disk that had about 24 small holes in a shallow spiral near the edge. The holes scanned a subject and separated the image into smaller sections. In one version, a lens focused reflected light from the subject onto photocells behind the rotating scanning disk. The cells translated the image into a pulsing electric current. The disk was named for Paul Nipkow (1860-1940) of Germany, who developed the idea in 1884. Nipkow's theory was sound, but no one had made such a device work. Sick, almost penniless, and with no laboratory equipment, Baird thought he could.

With his limited savings, Baird purchased some crude equipment and assembled it on a small table. It included a cracker box, a cardboard scanning disk, knitting needles, and army-surplus photocells and diodes. He used a Nipkow disk in his transmitter and receiver. The receiver recombined the scanned image onto a small screen. In 1924, Baird transmitted an image of a Maltese cross over a distance of a few feet. The lines of his picture went from top to bottom, not left and right as with current televisions. And the whole picture was made up of only 30 lines, not the 625 or more used in modern television sets. It flickered at 12 frames per second. His success encouraged Baird to take

The label accompanying this incomplete experimental mechanical television transmitter credited it to Paul Nipkow. Baird used the idea to make a practical system in 1925. The large holes in the spinning wheel reduce its weight. The scanning holes are in a shallow spiral and are too small to be visible in this photograph.

the apparatus to London, in hopes of finding financial backing.

Baird found a financial backer who offered a small amount of money and found him an upper-flat apartment that he could use for a laboratory. Almost always hungry, often sick, and frequently desperate, Baird spent months working on his crude equipment. His goal was to transmit a three-dimensional image and he used a puppet head as a subject. Success came in October 1925, when he transmitted a blurred but recognizable image of the puppet head. In January 1926, in his apartment-laboratory, he showed the transmission and reception of people's faces to an audience of technical people. The faces appeared as flickering pinkish images on a four-inch by two-inch screen. The size of the group may have been as high as 50. Baird became famous overnight. The publicity provided him with investment money and he was never poor again.

Like some others in technology, Baird was part successful inventor and part showman. He was eager to interest the public in television to create a market and established the Baird Television Development Company. He sent moving images along telephone lines from London to Glasgow in 1927, a distance of 483 miles. And in 1928, he stunned the world with a transatlantic signal from Purley, England, to Hartsdale, New York. The image was of a Mrs. Mia Howe and was received by amateur radio enthusiast Robert M. Hart.

The BBC granted Baird a six-year experimental license to transmit television signals using his mechanical scanning system. The station's call letters were 2TV and it went on the air in 1929. Television sets could be purchased in kit form for £12 or already wired for £20. Called Televisor, 10,000 to 20,000 sets may have been sold. At that time, programming in the traditional sense did not exist. The fuzzy images that people received were often only head and shoulder shots of people talking during test transmissions.

Baird had put up with many personal hardships and was unwilling to acknowledge the considerable benefits of all-electronic television. Some people thought him quite abrasive. He married Margaret Albu in 1931. She was a concert pianist and the daughter of a diamond merchant. She may have provided funding for some of Baird's later projects. The couple had two children. Baird died at the age of 58.

A major problem with Baird's system was that the rotating disk could only transmit low-definition images. Disks could not spin quickly enough to produce better results. The Televisor operated at only 30 lines of resolution and could not reproduce fine detail. It was for others to develop all-electronic television, which eliminated that problem.

But Baird's work was not a complete dead end. His approach received a new lease on life during the Apollo lunar landing missions of 1969-1972. The National Aeronautics and Space Administration (NASA) wanted to send live color television pictures from the moon. NASA awarded the camera contract to the Westinghouse Corporation. Existing color television cameras were bulky, heavy, and delicate. Westinghouse developed a 3-1/2-inch rotating-disk camera. It used red, green, and yellow filters that rotated at 600 rpm, the same speed used by Baird. The camera was used to televise all lunar landing missions.

References

John Logie Baird and Television by Michael Hallett, Priory Press, 1978.

Electrical Engineers and Workers by P. W. Kingsford, Edward Arnold Publishers, 1969.

The Communications Miracle by John Bray, Plenum Press, 1995.

"John L. Baird Dies; Television Leader," *The New York Times*, 15 June 1946, p. 21.

"The Color War Goes to the Moon" by Stanley Lebar, *American Heritage of Invention and Technology*, Summer 1997, pp. 52-54.

The 1929 Televisor had a rotating disk at the back. Its speed was calibrated with a knob at the left and the image was viewed through the small screen at the right.

Vladimir ZWORYKIN

Born:
July 30, 1889, in Murom, Russia

Died:
July 29, 1982, in Princeton, New Jersey

Television may currently be the world's most influential media form. It has been with us since the National Broadcasting Co. (NBC) established the first regular U.S. telecasts in 1939. Like many electronic innovations, television has no clear-cut single inventor. But one American stands out as the person whose achievements were pivotal to television's development. He held more than 120 patents. His best known are patents for the iconoscope camera tube (*icon* means "image") and the kinescope picture tube (*kine* means "motion"). Russian-born Vladimir Zworykin's contributions were so important that he is often described as the inventor of television—a title he always rejected.

Smithsonian Institution Photo No. 79-11567

Zworykin was born in a small town 200 miles east of Moscow near the Oka River. His father owned and operated a fleet of river boats, and the young Zworykin helped him during school vacations. He learned basic electricity on river journeys by reading books and making observations. He eagerly repaired electrical equipment. He obviously had greater interest in electricity than in shipping.

After graduating from high school in Murom in 1906, he decided to study electrical engineering at the St. Petersburg Institute of Technology. This new environment was quite a change for a young man raised in the countryside. St. Petersburg was the second-largest city in Russia, a cultural center, and about 700 miles from Zworykin's home. He almost immediately met Professor Boris von Rosing, who was working on transmitting pictures by wire. Von Rosing was the first person to attempt transmission of an image by scanning the inside of a cathode-ray tube. He freely allowed Zworykin to assist him in his research. He felt that the future of television lay in the cathode-ray tube—not in the mechanical systems being investigated by others. Zworykin spent much of his time with von Rosing blowing glass to form photocells and amplifying tubes. He learned much during his years at the institute and stayed on after graduation as an assistant for a few more months. Zworykin then went to Paris and assisted in X-ray experiments until the outbreak of World War I.

Zworykin made his way back to Russia and joined the Russian Signal Corps. He served until 1918, much of the time working on wireless transmission equipment near the Polish border. The Russian Revolution followed on the heels of the world war, plunging Zworykin's country into chaos. He wandered for months to avoid arrest by competing armies. Finally, he made his way to the northern port city of Archangel. Pleading his case to an American official, he received a visa. He sailed to London and then on to New York, arriving in the United States in 1919.

Zworykin first took work as a bookkeeper for the financial agent of the Russian Embassy in Washington, D.C. He moved to Westinghouse Electric and Manufacturing Co.'s research labs in Pittsburgh the following year. On his first assignment there, he worked on new radio tubes and photoelectric cells. He went to the University of Pittsburgh at night, earning a doctorate in 1926. He had become an American citizen in 1924, the same year that he first demonstrated a television system.

Zworykin had applied for a patent the previous year for his iconoscope, or television transmitting tube. He showed Westinghouse executives the first flickering images from his experimental system. His method differed from the cumbersome mechanical system of whirling perforated disks that had dominated early television development. Of the demonstration, Zworykin later said, "I was terribly excited and proud. After a few days I was informed, very politely, that my demonstration had been extremely interesting. But it might be better if I spent my time on something a little more useful." The same year, he patented his kinescope, or television receiving tube. The two components set the stage for a practical television transmitting and receiving system.

Unable to convince Westinghouse executives of the value of his inventions, Zworykin moved to the Radio Corporation of America (RCA) in 1929. He became director of its electronic research laboratory in Camden, New Jersey. Company president David Sarnoff encouraged Zworykin to develop the equipment necessary to make television practical. Himself a Russian immigrant from Minsk, Sarnoff provided Zworykin with everything he requested. In November, Zworykin demonstrated the first practical and completely electronic television system at a convention of the Institute of Radio Engineers. Over the next 20 years, RCA spent $50 million—a huge amount

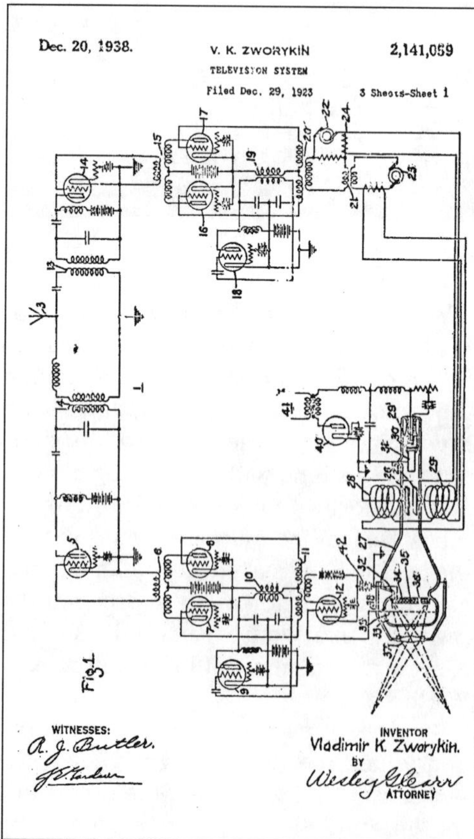

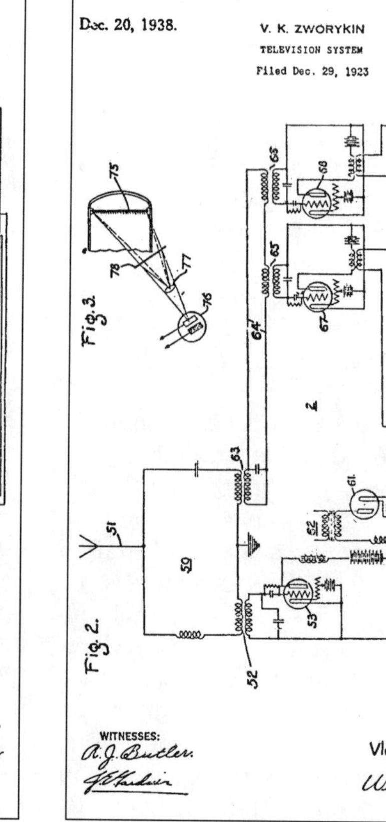

First two pages of drawings for Zworykin's television system patent. His design included both a transmitter (camera) and receiver (television). It was the first successful all-electronic system.

of money—perfecting a television system.

Zworykin and his staff worked diligently on the project. Their main concern was electronic sensitivity. They continually worked to produce more sensitive iconoscopes and kinescopes. For all its refinements, the modern television picture tube remains essentially the kinescope that Zworykin patented in 1924. RCA installed television sets in 150 New York City area homes in 1936 and began experimental telecasts. Felix the Cat, a popular cartoon character of the time, was the first image broadcast. NBC, a division of RCA, established regular telecasts in 1939.

Zworykin's other accomplishments included a color picture tube patent he took out in 1929. Using his knowledge of photoelectric tubes and image multipliers, he developed a rifle scope that allowed soldiers in World War II to see in the dark. Zworykin developed an electric eye that operated electric switches and automatically opened doors. He made vastly superior detector tubes for measuring radioactivity. He also worked on early computers. The photograph on the first page of this profile shows him holding an experimental data-storage tube.

After the television system, Zworykin's next-most-important work was with the electron microscope. The microscope was his idea, though he and James Hillier worked on it as a team during the early 1940s. Instead of focusing an image of reflected light with lenses, electron microscopes focus electrons with electromagnetic fields. Zworykin and Hillier reduced the microscope's size from an experimental device that almost filled two rooms to a 16" portable model. Electron microscopes magnified as much as 200,000 times, while optical microscopes had a limit of 2,500. Their improvement allowed researchers to identify such incredibly small things as metallic grain structure or viruses.

Zworykin retired from RCA in 1954, but he kept an office at the RCA laboratories. He never wanted to retire from technology. Zworykin was a humanist and was keenly aware that inventions do not automatically serve the public interest. He wanted to stay involved so that he could express his opinions. What free time he did have went to his favorite pastimes of swimming, tennis, and woodworking. He and his wife, Katherine, kept an open house for their 5 children and 17 grandchildren. The younger people passed in and out of the Zworykin's Princeton home with erratic frequency. Zworykin died of natural causes the day before he turned 93.

In 1966, President Lyndon Johnson awarded Zworykin the National Medal of Science, America's highest scientific honor. It was the most significant of the 27 major awards he received, which included several honorary doctorates. A modest person with a well-developed sense of humor, Zworykin coauthored many books. They included *Television* (1940) and *Electron Optics and the Electron Microscope* (1945). He once said that his response to the statement "It can't be done" would always be "Want to make a bet?"

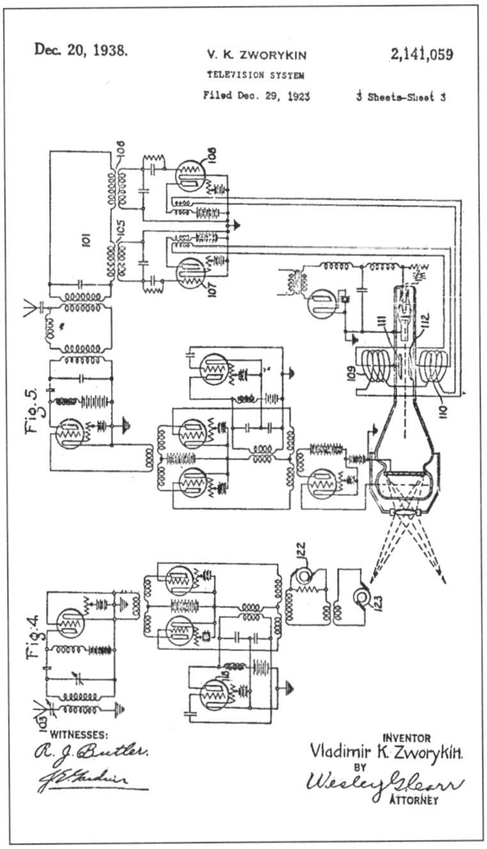

Dec. 20, 1938. V. K. ZWORYKIN 2,141,059
TELEVISION SYSTEM
Filed Dec. 29, 1923 3 Sheets-Sheet 3
Fig. 5
Fig. 4
WITNESSES: R. J. Butler.
INVENTOR Vladimir K. Zworykin.
BY Wesley Glenn ATTORNEY

Final page of the television system patent drawings

References

Those Inventive Americans, National Geographic Society, 1971.

American Science and Invention by Mitchell Wilson, Bonanza Books, 1960.

McGraw-Hill Encyclopedia of Biography, McGraw-Hill Publishers, 1973.

Asimov's Biographical Encyclopedia of Science and Technology by Isaac Asimov, Doubleday & Co. Publishers, 1964.

Vannevar BUSH

Born:
March 11, 1890, in Everett, Massachusetts

Died:
June 28, 1974, in Belmont, Massachusetts

In selecting a radio station, adjusting a microwave oven, or checking time on a wristwatch, people often read numbers from a digital display. By contrast, some radios, ovens, and watches have just a few numbers permanently in place on a dial face. With an *analog* wristwatch, for example, you determine time by the position of the two hands. While analog wristwatches use position to arrive at a time, digital watches use counting. Computers can also be classified as either digital or analog. Digital computers use input from keyboards. Analog computers operate with input of quantities of things such as voltage or speed. Heating-system thermostats and speedometers work on this principle. All analog computers descend directly from the first one built by the lanky Vannevar Bush in 1930.

Smithsonian Institution

The grandson of a whaling-ship captain and the son of a minister, Bush attended Tufts College, where he earned two degrees in engineering. He received his first patent while he was still in school. It was for a profile tracer, a surveying machine that measured elevations as a bicycle-tired device rolled along the ground. The machine included servomechanisms and other components that Bush would later use in his analog computer. Bush so impressed his instructors with his profile tracer that they granted him a college degree based on the invention's complexity. After graduation, he took a $14-per-week job testing machinery for the General Electric Co. in Schenectady, New York. After a suspicious fire destroyed some equipment, the company suspected foul play. It fired several employees, including Bush.

Bush served in the U.S. Navy during World War I, conducting research on submarine detection. He also worked at a few other jobs before moving to the Massachusetts Institute of Technology (MIT), where he spent most of his professional career. Besides teaching at MIT, Bush served as the institute's dean of engineering from 1932 to 1938. He also acted as a consultant to many companies. He was a brilliant investigator and organizer. He founded several successful companies, including, with two other persons, the Raytheon Corp., which now employs more than 76,000 persons.

Bush was particularly interested in problems associated with electrical power transmission. His investigations led him to think about the value of a calculating device that could solve complex problems. While working to eliminate blackouts, Bush managed to solve one particularly complex problem. It took him several months. To shorten the time required for necessary calculations, Bush invented his computer in 1930, calling it a *differential analyzer*. Because he was so busy at the time, he did not patent the computer. The paperwork involved seemed too troublesome.

Bush was a highly intelligent person who was rather colorfully described as a "reedy plain-spoken New Englander with a rustic grin and cracker-barrel drawl that concealed a mind of whiplash speed." With his expansive intellect, he could understand abstract mathematical concepts and visualize hardware to use those concepts. He took several years to invent his totally mechanical analog computer. It used motor-driven shafts and gears to multiply, divide, and solve complicated problems. The heart of his invention was a disk rotated by an electric motor. A wheel rolled on top of the disk in much the same way that a phonograph needle tracks the surface of a record. The wheel's speed changed as it moved closer to the center of the disk. Bush called the disk an integrator, and he used six of them in his analog computer. The photograph below shows one.

Long shafts interconnected six integrators and several small gear boxes. Like many other prototypes, the assembly was crude. It resembled something created using a huge Erector set. The photograph above shows Bush, at the far left, working on his computer in 1929. The other four men are working on the input/output tables, which resembled drafting tables. The computer was not easy to use and pro-

Smithsonian Institution Photo No. 58197

Bush (at left in photo) worked with several colleagues at MIT to build, test, and regularly operate his analog computer. Inputs and outputs used tracings on the tables in the photograph.

gramming it took up to two days. Three or four people were stationed at the tables. They continually observed and adjusted input pointers to keep them on track. At the output tables, other operators made notations and recorded the graphical results. The output was not digital—it appeared only as graphs or lines on a piece of paper. Interpreting the results took great skill.

Bush's analog computer was quite influential in the technical community. It offered an impressive demonstration of the computational power of machines. Because the computer was a mechanical device, it was not 100 percent accurate. It typically achieved 98 percent accuracy—a level that was quite acceptable for technical calculations in 1930. Not knowing exactly how to describe the computer to average people, some experts described it as a "mathematical robot."

The U.S. Ballistics Research Laboratory in Maryland and the University of Pennsylvania each wanted one of the computers. Contracts were signed and two 20'-long analog computers were completed in 1935. They were the first contract-built computers in the world. Much more elaborate computers were constructed at MIT over the next 15 years. At present, analog computers are used in flight simulators to teach pilots to fly airplanes, in hospitals to monitor patients, and in industrial process control.

Besides the analog computer, Bush also invented a network analyzer to test the ability of power systems to perform under heavy loads. He modified vacuum tubes and produced new gaseous conduction devices.

One of the six integrators used on Vannevar Bush's 1930 analog computer

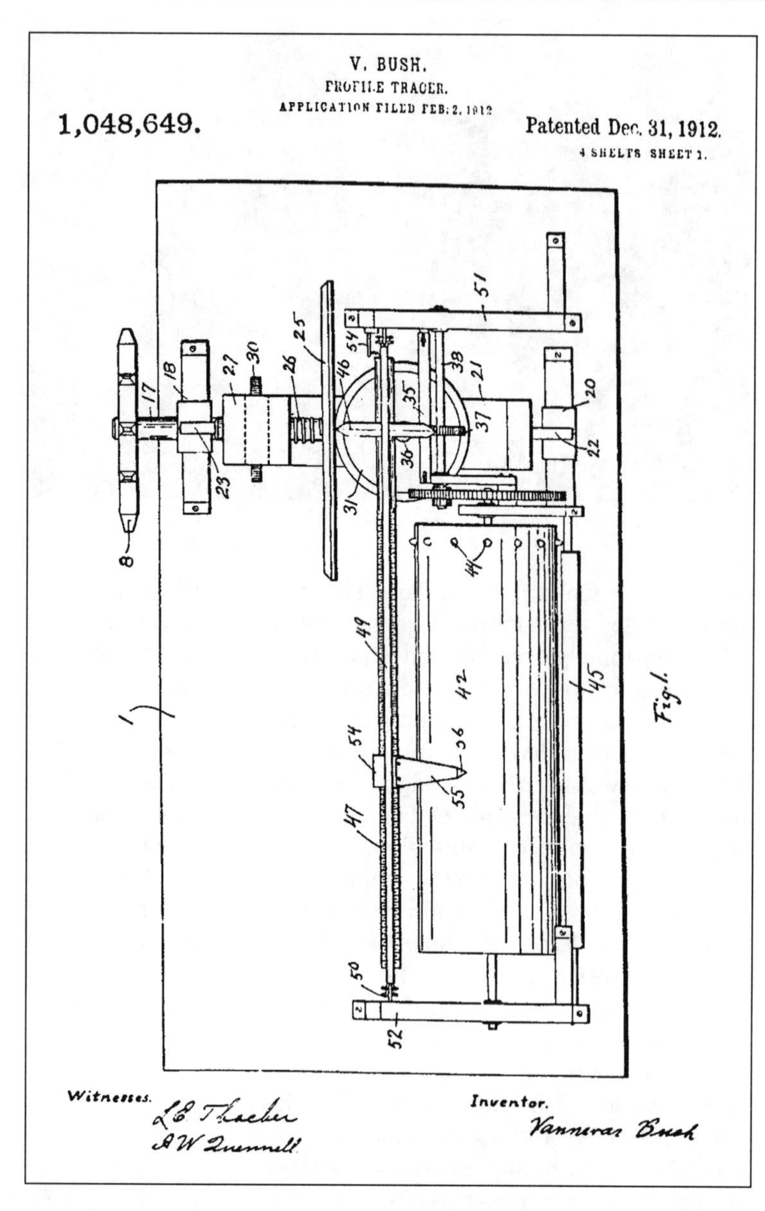

Bush's first patent was granted when he was in college. It measured ground elevations with a servo mechanism.

Once, when asked how many patents he had, Bush replied, "20 or 30."

Bush was an excellent organizer who served in several administrative capacities before, during, and after World War II. He was president of the multimillion-dollar Carnegie Foundation from 1939 to 1955. Between 1939 and 1941, he served as chair of the National Advisory Committee for Aeronautics, the predecessor to the National Aeronautics and Space Administration (NASA). He was also director of the Office of Scientific Research and Development from 1941 to 1946. In that position, he coordinated the wartime efforts of 25,000 science and technology workers involved with such complex projects as radar and atomic energy. After the war, he devoted his efforts to civilian control of atomic energy and to establishing the National Science Foundation in 1950.

There was also a light side to Bush's life, and he told some amusing stories in his 1970 autobiography, *Pieces of the Action*. One dealt with having to cope with driving an old Stanley Steamer motorcar on winter roads. In 1915, he was dating a young woman who lived at the top of a hill that often had an icy patch on the road. He said, "[I had to start at] the foot of the hill, look about for possible cops, pull the throttle way down, and roar over the patch." Bush wrote articles on many subjects, but his other three books dealt with his favorite topic, science and society. His hobbies included raising turkeys and playing the flute.

References

Pieces of the Action by Vannevar Bush, William Morrow Publishers, 1970.

The Computer from Pascal to Von Neumann by Herman H. Goldstine, Princeton University Press, 1972.

Standard and Poor's List of Corporations, 1990.

Asimov's Biographical Encyclopedia of Science and Technology by Isaac Asimov, Doubleday & Co. Publishers, 1964.

Natonal Cyclopedia of American Biography, James T. White and Co. Publishers, 1891; with supplemental updates.

Edwin ARMSTRONG

Born:
December 18, 1890, in New York, New York

Died:
February 1, 1954, in New York, New York

If radio communication has an unsung hero, it is certainly Edwin Howard Armstrong. He single-handedly developed the superheterodyne circuit, the basis for 98 percent of all modern radio and television receivers. Armstrong also developed many other useful circuits, but he is best known as the inventor of static-free FM radio. His *frequency modulation* principle is used not only in radios and televisions, but also in telephones, radar, and spacecraft communication networks. Armstrong lived an active life filled with discovery, invention, financial success, court battles, and, ultimately, tragedy.

Smithsonian Institution Photo No. 43614

During his teenage years in Yonkers, New York, Armstrong was fascinated by the new technology of wireless communication. After reading Guglielmo Marconi's book *The Boy's Book of Invention*, he decided to become an inventor in the field of radio. In his room—overlooking the Hudson River and filled with crystals, coils, condensers, and resistors—Armstrong busied himself experimenting with electrical circuits. While still a student at Yonkers High School, Armstrong built a 125-foot-high antenna on his family's lawn. He had no fear of heights, and he would often climb the tower to make adjustments or just for fun. After graduation, he commuted by motorcycle to Columbia University, where he eventually earned a degree in electrical engineering. One of his teachers was Michael Pupin, developer of the radio circuit that made possible tuning in just one radio station at a time.

Under Pupin's influence, Armstrong made his first significant discovery while he was still in college. While mountain climbing in Vermont in the summer of 1912, he devised a regenerative circuit for use with the new audion radio tube. Returning to complete his senior year, Armstrong built the circuit and found that it greatly improved radio reception. The circuit permitted hearing distant stations without the use of headphones. He filed for a patent, but World War I intervened before his circuit could gain wide acceptance.

Armstrong received a captain's commission in the U.S. Army Signal Corps and went to France. He worked on developing a system to detect enemy aircraft from the high-frequency pulses emitted by their spark plug firings. At first, the firing frequency was too high to permit easy reception. Armstrong developed a circuit that lowered the frequency and then amplified it. Using an antenna on the Eiffel Tower in Paris, his eight-tube superheterodyne circuit worked superbly. (*Heterodyne* refers to the mixing of radio signals.) Although Armstrong developed his system too late to play a role in the war, the superheterodyne circuit could be used in ordinary AM (*amplitude modulated*) radio receivers. When Armstrong returned home with the rank of major, the Institute of Radio Engineers (IRE) gave a dinner in his honor. The IRE awarded Armstrong its first Medal of Honor. The medal recognized his status as the foremost expert in his field.

Armstrong was known as Howard to his friends, and in adult life he was almost totally bald. He lost his hair following an illness during his overseas wartime service. After the war, he continued to work at a Columbia University laboratory on circuits that improved the sensitivity of radio receivers. His work brought him in contact with the Radio Corporation of America (RCA), which paid him quite well for his patent rights. He became a millionaire during the 1920s. He also dated the secretary of RCA's president. It may have been the young woman, Marion MacInnis, who inspired Armstrong to scale the WJZ transmitting tower, 450 feet above New York's 42nd Street, in May 1923. He posed for several photographs, without safety gear. That stunt got him barred from the RCA offices for several weeks, but he married MacInnis the following December. His wedding present to her was a suitcase-sized portable superheterodyne radio, the first ever made. The couple drove Armstrong's new Hispano-Suiza convertible to Florida for a honeymoon and lugged the radio to the beach for the benefit of reporters and photographers.

Edwin Armstrong demonstrates his FM radio system in 1939. William Baker (rear) was head of radio research at General Electric.

Hall of History Foundation, Schenectady, NY

Armstrong continued to work on FM radio in his laboratory. As the decade wore on, he found himself trapped in a corporate war to control radio patents. RCA, Westinghouse, American Telephone & Telegraph, and others all aimed to build large corporations based on the patents of different individuals. With 60,000 homes using radios in 1922 and projections for rapid growth, there was a great deal of money at stake. The number of homes with radios boomed to 2.75 million in 1925 and 14 million in 1930. Armstrong became involved in complex court battles that dragged on for years. In 1934, after 12 years of litigation in one particularly important suit, the U.S. Supreme Court handed down a verdict against Armstrong. It was a heavy blow to him and his supporters. Many people felt the members of the Supreme Court based their judgment on a misunderstanding of the technical facts. Armstrong offered to return his Medal of Honor to the IRE but the institute refused to accept it. In a further show of support, the Franklin Institute weighed all the technical evidence and awarded Armstrong the highest honor in U.S. science, the Franklin Medal. Throughout all the years of testimony, he continued his work to eliminate radio static.

During the early 1930s, practically everyone thought that frequency modulation was useless for communication. Nonetheless, Armstrong felt that FM was the only solution. He had already decided there was no way to eliminate static from conventional AM radio transmission. Since RCA had heavy investments in AM transmitters and receivers, company officials finally asked Armstrong to remove his equipment from the space RCA provided for him in the Empire State Building. He moved to a large apartment overlooking the East River and personally financed his continuing experiments with FM. He field tested his efforts in 1933 during a violent thunderstorm and received static-free high-fidelity sound from

80 miles away. On July 18, 1939, Armstrong began broadcasting from the world's first FM station, W2XMN, which he had built in Alpine, New Jersey, with his own money.

Once again, a war interrupted Armstrong's work. He spent World War II working on radar. After the war, RCA brought out its first FM receiver, which supposedly used a new circuit to eliminate static. The circuit was effective—and it was obviously an adaptation of Armstrong's patented circuit. Armstrong filed a suit against RCA. Although he was clearly in the right, he knew he couldn't win the legal battle. Regarding the situation, this man of few words said, "They will stall this along until I am dead or broke." RCA did stall, and Armstrong's legal fees steadily mounted. The financial and emotional strain proved more than he could bear. Armstrong took his own life in 1954.

Shortly after her husband's death, Armstrong's widow won $10 million in damages in 21 patent-infringement suits. The proceedings weren't completely closed until 1967, when the Supreme Court refused to review a lower court judgment against Motorola.

The International Telecommunications Union in Geneva, Switzerland, elected Armstrong to its roster of communication pioneers. His name appears with Marconi, Pupin, and Alexander Graham Bell. Today, there are about 5,000 FM stations in the U.S. All are testimony to the creative genius of Howard Armstrong.

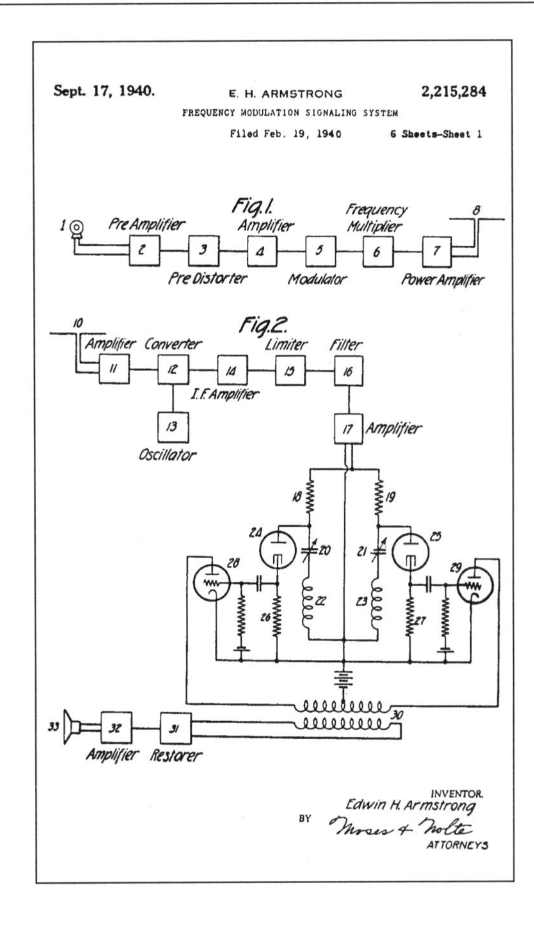

One of Armstrong's FM patents used simple block diagrams.

References

Empire of the Air by Tom Lewis, Edward Burlingame Books, 1991.

The Smithsonian Book of Invention, Smithsonian Institution, 1978.

"Radio Revolutionary" by Thomas S. W. Lewis, in *American Heritage of Invention and Technology*, Fall 1985.

Dictionary of American Biography, Charles Scribner's Sons Publishers, 1932; with supplemental updates.

McGraw-Hill Encyclopedia of Biography, McGraw-Hill Publishers, 1973.

Asimov's Biographical Encyclopedia of Science and Technology by Isaac Asimov, Doubleday & Co. Publishers, 1964.

Leopold MANNES AND *Leopold* GODOWSKY II

Leopold Mannes
Born:
December 26, 1899, in New York, New York

Died:
August 11, 1964, on Martha's Vineyard, New York

Leopold Godowsky
Born:
May 27, 1900, in New York, New York

Died:
February 18, 1983, in New York, New York

Career technologists have not been the only ones to improve products and processes. A pair of classical musicians combined their technical talents in 1935 to develop the world's most popular color-slide film. American-born Leopold Mannes and Leopold Godowsky invented Kodachrome film at the Eastman Kodak Co. research laboratory in Rochester, New York.

The search for color-sensitive emulsion began during the 19th century. Early investigators took photographs of the same scene through colored filters, then projected superimposed images onto a wall through complementary filters. The projected image was a color rendition that used an additive process, because filters added the color. The emulsion that Mannes and Godowsky perfected used a subtractive process. It was based on colored objects absorbing—or subtracting—some light rays and reflecting others.

Both Mannes and Godowsky were born into families with strong musical backgrounds. Mannes's uncle Frank Damrosch founded the Julliard School of Music, and Godowsky's father was a world-class pianist and composer. The two met at age 15 while attending the Riverdale Country School in New York City. They initially noticed each other because they shared the same first name. They soon discovered that they also shared interests in music, amateur chemistry, and photography. Mannes was to become a famous pianist and Godowsky would achieve equal fame with the violin. During their teen years, however, they were fascinated by attempts to devise a practical method for taking color photographs. Their meeting sparked a friendship and professional working relationship that lasted their entire lives.

Leopold Mannes (left) and Leopold Godowsky in their research lab at Kodak in 1932

Mannes and Godowsky conducted chemistry experiments in the bathrooms, kitchens, and pantries of their parents' New York City homes. The two friends did not know that hundreds of prominent scientists had been searching for a practical color process for some 50 years. Godowsky later said, "We were blissfully ignorant. Our physics instructor gave us a passkey to the [chemistry] laboratory and we spent a lot of time experimenting, boy fashion."

After graduating from high school, Mannes attended Harvard University and earned a degree in music. He also took many physics classes. Godowsky majored in chemistry and physics on the other side of the country, at the University of California. He started in Berkeley, where he played violin with the San Francisco Symphony Orchestra. He graduated from UCLA, where he played with the Los Angeles Philharmonic Orchestra.

After college, the two men began separate musical careers in New York City. Mannes taught music at the David Mannes College of Music, founded by his father and mother in 1916. The younger Mannes had a real talent for chamber music, and he received a Pulitzer Prize in music composition in 1925. Godowsky played with orchestras and continued studying music composition.

Mannes and Godowsky worked on the color-film project in their spare time. After several years of effort, they narrowed their investigations to an integral tripack: three layers of photographic emulsions on the same film base. The emulsion in contact with the celluloid film base was sensitive to red light. The next layer was sensitive to green light, and the top layer was sensitive to blue. To protect their interests, Mannes and Godowsky applied for patents. They received their first in 1924. Their experiments expanded, and they had to rent laboratory space. They paid for rent and supplies out of their earnings as musicians, but they soon realized that they would need financial assistance. In the late 1920s, they approached Kenneth Mees, a vice-president of Eastman Kodak and a strong supporter of photographic research. Mees agreed to provide some assistance in exchange for special considerations if they succeeded.

When George Eastman, the founder of Eastman Kodak, heard of Mannes and Godowsky's work, he wanted to buy the rights to their early patents. Eastman had a policy of paying a fair price for all patents dealing with photography. (Twin brothers F. E. and F. O. Stanley, for example, sold Eastman their patents for photographic dry-plate manufacturing machines. They used the proceeds to start their Stanley Steamer Motor Car Co.)

Because of economic hardship caused by the Great Depression, Eastman persuaded Mannes and Godowsky to join his company's research staff. He immediately paid each one $30,000 and offered an annual salary of $7,500. They started working in 1931 at a new research building with 12 technical assistants.

To measure critical development times in the complete darkness of a color darkroom, Mannes and Godowsky whistled the last movement of a classical music piece that they both knew well. Strains of the *C-Minor Symphony* by Johannes Brahms, at a precise two beats per second, came from their darkroom almost every day. Their coworkers were not accustomed to such unusual lab activities, but the two Leos—as their friends called them—quickly earned everyone's respect. They accumulated a total of 40 patents, but Kodachrome remains their best-known invention. It was rated at

Diagram for film requiring dye toners

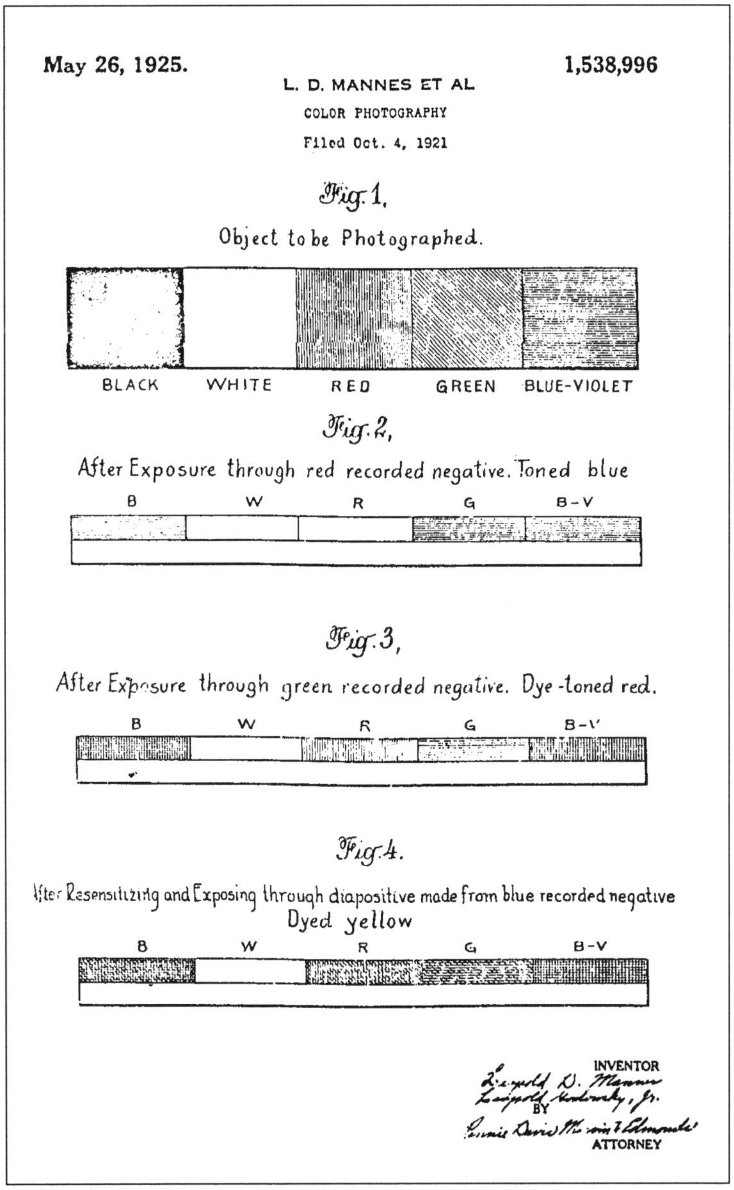

May 26, 1925. 1,538,996
L. D. MANNES ET AL
COLOR PHOTOGRAPHY
Filed Oct. 4, 1921

Fig. 1.
Object to be Photographed.
BLACK WHITE RED GREEN BLUE-VIOLET

Fig. 2.
After Exposure through red recorded negative. Toned blue
B W R G B-V

Fig. 3.
After Exposure through green recorded negative. Dye-toned red.
B W R G B-V

Fig. 4.
After Resensitizing and Exposing through diapositive made from blue recorded negative
Dyed yellow
B W R G B-V

INVENTOR
BY
ATTORNEY

a slow ASA (ISO) 10 when it first went on sale as 16 mm movie film on April 15, 1935. The following year, Eastman Kodak made it available for 35 mm still cameras. Kodachrome is still highly regarded for its sharpness, brilliance of color, lack of grain, and minimal color shift over time. Well-stored Kodachrome transparencies from the 1930s retain their beautiful color today.

Kodachrome is unique in that the colors seen after development are not part of the original tripack film. Kodachrome is the only color film that is black and white when exposed. Colors are added during development by a process called *dye coupling*. A dye coupler is a chemical that will react with a developed image to form a specific color. The first step is developing the film to obtain a negative, and then chemically reversing it to a positive. During three additional developments and color reversals, complementary colors of yellow, cyan (a greenish blue), and magenta (a purplish red) are introduced to the appropriate film layer. Silver is bleached away, leaving a color transparency. Amateur photographers cannot develop Kodachrome because it requires complex processing and quality control. In the late 1980s, only three dozen laboratories in the world could develop it.

Mannes and Godowsky left Kodak to return to concert music. However, the more inventive Mannes stayed with Kodak as a part-time consultant. He patented a sound track of gold in 1941 that improved the sound quality of color motion pictures. He also returned to teaching at his parents' music college.

Godowsky was a member of several prominent symphony orchestras, and he returned to professional violin playing. His father had been such a great pianist that the younger Godowsky took up the violin to escape his father's shadow. He even carried the influence of music into marriage. He married Frances Gershwin, the younger sister of George Gershwin, who composed such popular orchestra music as *Rhapsody in Blue*.

The two Leos were well rewarded over the years with royalties and the admiration of their peers. Almost 30 years after Kodachrome went commercial, international photographers had not forgotten Mannes and Godowsky. The two men received the Progress Award of London's Royal Photographic Society in 1964, the year that Mannes died. A reception at the George Eastman House in 1985 honored the fiftieth anniversary of Kodachrome's introduction to the public. At the opening, the Canterbury Brass Quintet performed separate musical pieces written by George Gershwin and Leopold Godowsky. After five decades of use, Kodachrome is a rare technical product in that it remains essentially unchanged.

References

The History of Photography by Beaumont Newhall, Museum of Modern Art, New York, 1964.

"Time Exposure" by Eaton S. Lothrop, Jr., in *Popular Photography*, February 1986 .

"Kodachrome Still a Leader in Color Film," in *Lexington (Kentucky) Herald-Leader*, 7 July 1985.

Dictionary of American Biography, Charles Scribner's Sons Publishers, 1932; with supplemental updates.

National Cyclopedia of American Biography, James T. White and Co. Publishers, 1891; with supplemental updates.

Howard AIKEN

Born:
March 9, 1900, in Hoboken, New Jersey

Died:
March 14, 1973, in St. Louis, Missouri

Computers are such a common part of modern technology that it is hard to imagine the world without them. When Charles Babbage (1791-1871) worked on his computer in Britain in the 1840s, no one understood what he was doing. Very few people could comprehend the concept of a nonhuman computer. Babbage began constructing what he called an *analytical engine*. The high cost of its many complex parts kept him from completing it. The British government had chosen not to get involved with the project.

By the 1930s, the world was considerably different. People flew through the air. Unseen electrons lighted houses with electricity. International news was immediately available through radio broadcasts. Technology's image had progressed to the point where average people generally understood the idea of technical development. But some technological developments did not benefit humanity. Europe and Asia were in political turmoil, with war clearly on the horizon. Howard Aiken had an idea for a computer that could help strengthen America. In the late 1930s, he convinced the International Business Machines Corporation (IBM) to develop his design, which he called the Mark I. Aiken's electromechanical computer was the world's first large-scale, program-controlled, general purpose digital computer.

Although born in Hoboken, Aiken grew up mostly in Indianapolis, where he attended Arsenal Technical High School. He worked 12 hours each night at an electrical power plant to help support his family and save money for college. He earned an electrical engineering degree from the University of Wisconsin in 1923. Aiken then took a job with the Madison (Wisconsin) Gas and Electric Company. He worked for two more companies before deciding to return to school for a physics degree. He started at the University of Chicago in 1933, then transferred to Harvard University. He received a Ph.D. in 1939 and stayed on at Harvard as a professor.

Smithsonian Institution Photo No. 81-10040

Aiken often worked with lengthy mathematical equations. The tedious work made him think of using a machine for routine and labor-intensive calculations. He used the writings of Babbage, the 19th century computer pioneer, as a basis for his theoretical automatic calculator. On paper, he devised a method in 1937 that used the on/off character of electrical relays to identify

Smithsonian Institution Photo No. 74-7494

The huge 50-foot-long Mark I computer was slow by modern standards and only had the calculating power of a modern $5 calculator. Four output electric typewriters are at the far right. Four paper-tape input devices are next to them.

numbers in binary code. His insight paralleled that of German computer pioneer Konrad Zuse (1910-1995). Neither knew about the work of the other. Aiken discussed his idea with colleagues, but they considered work on his hypothetical device too expensive to be sponsored by Harvard. They suggested that Aiken contact IBM. At that time, the company manufactured calculators, accounting machines, printing tabulators, and other office equipment.

In late 1937, Aiken met with James Bryce, one of IBM's most respected inventors, who held 500 patents. His most significant inventions were the multiplying and dividing mechanisms used in calculators. Both would be important components in the Mark I. The clarity of Aiken's proposal and its potential for success impressed Bryce. After several meetings, the two men approached company president Thomas Watson (1874-1956). They were surprised at how rapidly Watson agreed to support the expensive venture. Watson often made quick decisions. But he may have also been influenced by the possibility of a second world war, which would begin in 1939. Ultimately, IBM paid two-thirds of the $500,000 project cost and the U.S. government paid the rest.

Work began in 1939 at IBM headquarters in Endicott, New York. Aiken supplied the insights and IBM engineer Clare D. Lake assembled the hardware with a small group of assistants. Lake's many inventions over 30 years with IBM included the company's first printing tabulator. His successes earned him the nickname "Mr. Accounting Machine." Although Aiken's invention would eventually become a general-purpose computer, IBM called it the Automatic Sequence Controlled Calculator. Everyone else called it the Mark I.

Input information was delivered to the Mark I by four punched paper-tape readers.

Aiken tried to make the machine as simple as possible. It addressed the four operations he considered essential. He thought that the computer should (1) use positive and negative numbers, (2) use mathematical functions like sines and tangents, (3) operate automatically, and (4) use normal mathematical sequences. First run in January 1943, the Mark I was huge: more than 50 feet long, 8 feet high, and 5 feet wide. Its 750,000 parts included counters, punches, cam contacts, clutches, and countless rotating shafts. It used four paper-tape readers for input information. One carried the program instructions and three held the data. Four electric typewriters recorded the output. More than 500 miles of wiring connected everything. The key items were the 3,000 relays that supported the binary mathematics necessary to efficiently carry out calculations. The noisily clicking relays made the operating Mark I sound like a room full of people knitting. The computer was highly secret during the war.

After a brief test run in Endicott, the Mark

Smithsonian Institution Photo No. 55754

I was disassembled and shipped to Harvard University near Boston. By this time, Aiken had been drafted into the service for wartime duty. As Lieutenant Aiken, he supervised the Mark I's reassembly and its full-time operation for the Navy. Programming the Mark I required adjusting as many as 1,400 rotary switches on an external panel. Long sections of three-inch-wide paper were punched with programming holes and fed through one of the four tape readers. Lieutenant Grace Hopper (1906-1992) was one of the computer's three programmers. She and Aiken worked together for six years. Hopper later helped to develop the popular Common Business Oriented Language (COBOL). In 1983, she became the first woman to achieve the rank of Rear Admiral.

The Mark I took 0.3 seconds to add or subtract, 4 seconds to multiply, and 12 seconds to divide. Four people once took three weeks to solve a problem that the Mark I finished in only 19 hours. It operated around the clock and was mainly used to calculate trajectories for shells fired from large navy guns. When journalists described the Mark I to the public after the war's end in 1945, they often incorrectly labeled it an "electronic brain." It was electromechanical and had no electronic components. Aiken shared a post-war patent for it with three IBM employees. With the complex nature of the patent and the growth of competing methods, it took more than seven years between the application for and the granting of the patent.

The Mark I operated for more than 15 years until it was retired in 1959. The machine was carefully dismantled with pieces going to IBM, Harvard, and the Smithsonian Institution. Although slow by modern standards, the Mark I was the first of its breed and launched America's leadership in the computer industry.

Aiken completed an improved version, the Mark II, in 1946. The electronic Mark III was operational in 1950, and the Mark IV in 1952. After that, Aiken got out of the business of building computers. He became director of a new computer facility that he founded at Harvard. His greatest achievement might have been helping universities establish computer science programs. He was a quick-tempered person and difficult to get along with. Due to a conflict with

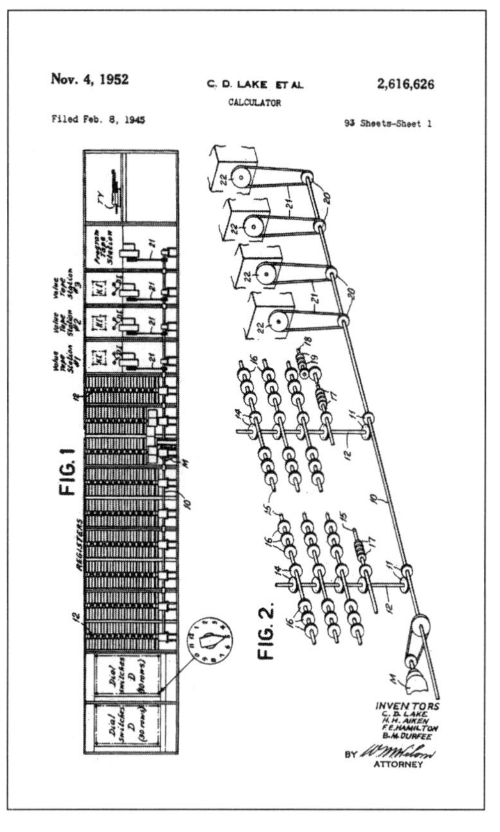

Aiken's Mark I patent consists of a large stack of paper: 93 pages of drawings and 138 pages of text. IBM employee Clare Lake was listed as the primary inventor, in part because Aiken was so hard to get along with.

Thomas Watson, Aiken refused to acknowledge the IBM president's role in the Mark I project. But Aiken also had softer edges to his personality. He was not secretive about his work and openly shared his ideas with colleagues. He gave all lecture fees that he received to members of his staff. They often appeared as wedding gifts or loans between paychecks, which never had to be repaid.

Aiken retired to Florida in 1961 to do independent computer consulting, and he became a professor at the University of Miami. He received countless awards from professional organizations, the U.S. Navy and Air Force, foreign governments, and universities. He was a rare person whose insights have had a significant effect on the entire computing profession. Aiken died at the age of 73.

The Mark I showed the way for others as the 1940s brought a rapid increase in computer development. It is not unusual to read that the 1946 Electronic Numerical Integrator and Calculator (ENIAC) was the first computer. The ENIAC may have a claim as the first *electronic* computer, but the electromechanical Mark I had been operational three years earlier. Casually written books sometimes completely neglect the Mark I, but serious references never do. A typical example comes from *Portraits in Silicon* by Robert Slater (1989): "Aiken had managed to build the first program-controlled computer." That simple sentence says it all.

References

Portraits in Silicon by Robert Slater, MIT Press, 1989.

Encyclopedia of Computer Science and Engineering edited by Anthony Ralston & Edwin D. Reilly, Van Nostrand Reinhold, 1993.

The Making of the Micro by Christopher Evans, Van Nostrand Reinhold, 1981.

Timetable of Technology edited by Patrick Harpur, Hearst Books, 1983.

Journey through Inventions by Ron Taylor, Smithmark Publishers, 1991.

Timetable of History—Science and Innovation, Xiphias Software (CD-ROM), 1993 (address: 8758 Venice Blvd., Los Angeles, CA 90034).

Margaret BOURKE-WHITE

Courtesy of Syracuse University with permission of the Margaret Bourke-White Estate

Born:
June 14, 1904, in New York, New York

Died:
August 27, 1971, in Stamford, Connecticut

Most factory workers of the early 20th century did not notice the beauty hidden in their industrial surroundings. Their workplaces were so familiar that most took no special notice of the dynamic visual impact of flowing steel. Few saw the symmetry in stacked airplane wings or the stark comparisons between people and the huge pieces of equipment they built. It took the trained and insightful eye of Margaret Bourke-White, one of the first industrial photographers, to isolate detail and record the visual drama of America at work.

This tall and strikingly handsome woman chose to add her mother's family name, Bourke, to her own while she was in her early twenties. Bourke-White's mother was a demanding person who expected her daughter to take the more challenging way to solve any problem. Her father was a would-be inventor and engineer in the printing industry. He once described his philosophy as "never leave a job until you have done it to suit yourself and better than anyone else requires you to do it."

Bourke-White graduated from Plainfield (New Jersey) High School and attended colleges in Michigan, Indiana, and Ohio. She earned a degree in biology in 1927 from Cornell in New York state. Bourke-White had a lifelong love of animals and expected to work with reptiles after college. She would occasionally take a pet snake with her to her classes. By the time she attended Cornell, her father had died, and to support herself she photographed buildings on campus. She used a 3-1/4" x 4-1/4" Ica Reflex camera that had a crack right through its lens and cost $20. She sold prints she made for income. That was the casual beginning of a lifelong career that would see Bourke-White never more than an arm's length away from a camera. She once said that if she hadn't had to work her way through college, she would never have been a photographer.

After graduating, Bourke-White moved to Cleveland to try her luck with professional photography. Fascinated by the city's manufacturing strength, she photographed factory buildings and other parts of Cleveland's in-

This image of a 200-pound ladle was one that Bourke-White photographed at the Otis Steel Co.

dustrial community. She was not the first person to take such photographs, but she is regarded as the first to avoid static representations of form. She gave her images a moving dramatic character. Bourke-White made machinery look so beautiful that one of her peers said she "transformed the American factory into a Gothic cathedral." A display of her prints at a local bank brought her to the attention of city leaders and resulted in a variety of new assignments. Her first serious industrial photographs were taken over a five-month period at the Otis Steel Corp.

At Otis, Bourke-White experimented with lenses, films, and magnesium flares to provide illumination for her slow ASA (ISO) 12 film. She produced a series of impressive images under challenging conditions. Although she took over 1,000 pictures with a large tripod-mounted camera, she only presented her best 12. The company president liked the photographs and used them in a limited production magazine called *The Romance of Steel*. A few months later, the Associated Press ran a national article about Bourke-White's images. She was only 23, and the article was headlined: "Girl's Photographs of Steel Manufacture Hailed as New Art."

One of Bourke-White's technical contributions was the perfection of multiple-flash photography. Flashbulbs were not yet available so she burned magnesium in open pans. Her multiple-flash technique eliminated harsh high-contrast images such as those produced by a single flash. The magnesium pans had to be carefully positioned so that assistants could ignite them without showing up in the photograph. Synchronizing the lighting of several flashes at once proved a real challenge. Preferring cameras larger than 35 mm, Bourke-White's favorite was a 5" × 7" Corona View with Bausch & Lomb lenses. She used it to photograph miners, grain elevators, meat packing plants, oil refineries, electrical alternators, paper mills, watch factories, and countless other industrial subjects. People were routinely surprised to discover that the photographer was an attractive and fashionably dressed woman in her 20s.

A perfectionist, Bourke-White always spent a great deal of time setting up her shots. It was not unusual for her to place herself in life-threatening locations to obtain the best image. She moved in so close to the molten metal in a steel factory that the varnish blistered on her camera. Of one photograph, she wrote in her diary, "I am glad [it] is good because it was so exciting to go up and take it through the carbon monoxide gas on the top of the coke oven, with my guide posted at the foot of the steps to run up and catch me if I should keel over." The photograph on the first page of this profile shows the 25-year-old Bourke-White with an early reflex camera. For dramatic effect, she positioned herself on the outside of the Chrysler building's sixty-first floor during construction of the building. (Years later, she had a studio located on that very floor.)

Bourke-White's reputation grew to the point where *Fortune* magazine offered her a position as its first staff photographer. She was the only one to receive a credit line in the first issue, February 1930. The finest photographic magazine in the country at the time, *Fortune* emphasized manufacturing. Bourke-White, its premier photographer, grew synonymous in the public's mind with industrial photography. Bourke-White also worked as one of *Life* magazine's original photographers and was responsible for both the cover picture and lead story in the magazine's first issue of November 23, 1936. Consistent with her industrial photography background, the cover image showed three huge concrete dam supports. Part of the $100 million Fort Peck Dam on the upper reaches of the Missouri River in Montana, each was about 100 feet tall. When she left *Life* more than 20 years later, Bourke-White had completed 284 assignments.

Early in her career, Bourke-White's images emphasized machinery, while the worker played a secondary role. At the time, this photographic style appealed to the public. By the mid-1930s, however, the Great Depression forced the country to pay closer attention to the problems of human beings. More and more, Bourke-White focused her lens on the workers. Then she slowly began to lean toward photojournalism. In *Life*'s first issue, her cover photograph clearly demonstrated the technical might of America. However, practically all of her 16 photographs inside the magazine were of the *people* who were constructing the dam that appeared on the cover.

Bourke-White was a tireless and energetic worker who traveled with several assistants. She took 3,000 large-format photographs during several trips to Russia between 1930 and 1932. She was the first person to do a full documentary on that country's emerging technological strength. Her first book was the 1931 *Eyes on Russia*. It featured 40 of her pictures. Bourke-White was an excellent photographer of people. She demonstrated her talent by recording dramatically emotional images during World War II and the Korean War. "Maggie" to her closest friends, Bourke-White was briefly married two times. Her second marriage was to *Tobacco Road* author Erskine Caldwell. She died at her home in Stamford, Connecticut.

References

Portrait of Myself by Margaret Bourke-White, Simon and Schuster, 1963.

"Woman of Steel" by Vicki Goldberg, *American Heritage of Invention and Technology*, Spring 1987.

"Unforgettable Margaret Bourke-White" by Carl Mydans, *Reader's Digest*, August 1972.

Girls Who Became Artists by Winifred and Frances Kirkland, Books for Libraries Press, 1934.

Chester CARLSON

Born: February 8, 1906, in Seattle, Washington

Died: September 18, 1968, in New York, New York

It seems hard to believe that people did not always think that many copies of documents were necessary. Today, there are more than 5 million copiers in use in America, which turn out an estimated 2,000 copies each year for every person in the country. In the 1930s, people didn't see a need for multiple copies, and it took more than 20 years for the first copier to reach production. Chester Floyd Carlson invented his copier in 1938. He was the last person who single-handedly developed a new product that spawned an entire industry. The first practical dry copier, the Xerox 914, was introduced to the public in 1960.

Carlson was the only son in a poor and sickly family. His mother's tuberculosis led the family to move often in search of a more healthful climate. Because he moved so often, Carlson developed very few friendships as a child. His father was a barber who became bedridden with a spinal disorder. At 14, Carlson provided his family's primary financial support. He rose before sunrise at their San Bernardino, California, home to wash windows and clean businesses. He earned $50 to $60 per month. He also worked summers on a farm and in a cement plant. Carlson later said that his early experience introduced him to the value of work and the discipline that goes with it.

Showing impressive determination not to give up, Carlson put himself through a local community college. There, he met an engineering teacher who encouraged him to pursue photography and such outdoor activities as hiking and fishing. The teacher recognized Carlson as a brilliant student and pushed him to raise his goals. Carlson eventually graduated from the California Institute of Technology. He remained a lifelong friend of the teacher who inspired him.

Courtesy of the Xerox Corp.

In this 1965 photograph, Chester Carlson shows the materials he used in 1938 to make the first xerographic copy.

Following graduation from Cal Tech, Carlson landed a $35-per-week position with Bell Telephone Laboratories across the country in New York City. He worked in Bell's patent department until he was laid off in 1933, a victim of the Great Depression. He found another job with P. R. Mallory Co., a manufacturer of electrical components. This was the most significant move in Carlson's life. At Mallory, he worked long hours checking and comparing patent drawings and text. At the time, the only two methods available for duplicating such complex documents were photography or redrawing and retyping. Carlson decided there had to be a better way.

Carlson had owned a printing press during his high school years and had a continuing interest in the graphic arts. He began his work on the dry copier by searching the technical literature. He found nothing that suggested that others were working on such a duplication process. However, Carlson did find information about photoconductivity. Some materials, such as sulfur, change electrical conductivity after exposure to light. Experiments that Carlson conducted in his kitchen produced unpleasant odors that spread throughout the apartment house where he lived. The daughter of the apartment's owner came to complain, but instead she became interested in Carlson and his work. They married in 1934.

Carlson needed more space to continue his nighttime experiments. He rented a small room at the back of a beauty shop operated by his mother-in-law in Astoria, New York. Although he could hardly afford it, he also hired an assistant, a German physicist named Otto Kornei who had just immigrated to the U.S. Carlson budgeted $10 per month for his research. Carlson and Kornei made their first successful experiment in 1938. Kornei darkened a room and rubbed a sulfur-coated zinc plate with a handkerchief to develop static electricity. He pressed the sulfur plate against a glass plate that had words written on it and then exposed it to a bright light for about three seconds. Carlson then dusted the sulfur with yellowish lycopodium powder, a natural spore also known as club moss. He gently blew on the plate, removing the loose powder and leaving a temporary image. To make the image permanent, Carlson placed a sheet of waxed paper on the powder and heated it. The waxed paper held the world's first legible dry-copied image: "10-22-38 ASTORIA." Almost in disbelief, the men repeated the experiment several times. It succeeded equally well each time. To celebrate, the men put away the sandwiches they had brought with them and went to lunch at a nearby restaurant.

Carlson's patent was the first in the field. He called the process *electrophotography*. Since no one else was working on dry copiers, his patent was very broad based and gave Carlson many rights. Unsure of the project's direction, Kornei left six months later to take a corporate job. However, Carlson never forgot Kornei's help and generously rewarded him in later years with Xerox Corp. stock.

Carlson tried for five years to make a simple copying machine. He looked for a corporate sponsor, but more than 20 companies rejected his prototype. He was nearly broke in 1944 when an employee of Battelle Memorial Institute came to the Mallory Co. to discuss other patents. Battelle was a small research organization in Columbus, Ohio. Carlson engaged the Battelle representative in conversation, and he agreed to have technicians study Carlson's invention.

Battelle agreed to take over some of the design work, but soon spent the small $3,000 budget assigned to it. Batelle, too, went looking for corporate sponsors. Batelle met with rejection from everyone except the Haloid Corp., a manufacturer of photographic paper and other items in Rochester, New York. An agreement was reached in 1946 that gave Haloid a license to develop a copying machine based on Carlson's patents. The company figured that it would have to spend $25,000 per year in research on the device. Since it had earned only $101,000 in 1946, Haloid was taking a tremendous gamble.

Haloid introduced a crude and cumbersome copying machine in 1949 called the Xerox Model A. Its operation required 14 separate steps, and business and industry leaders did not accept it. The term *xerographic* comes from the Greek xeros, which means "dry," and *graphos*, which means "writing." The word was first used in 1947. Dozens of difficult technical problems with the copier awaited solution. The company experimented with other dry models be-

Courtesy of the Xerox Corp.

The 1960 Xerox 914 was the world's first practical dry copier.

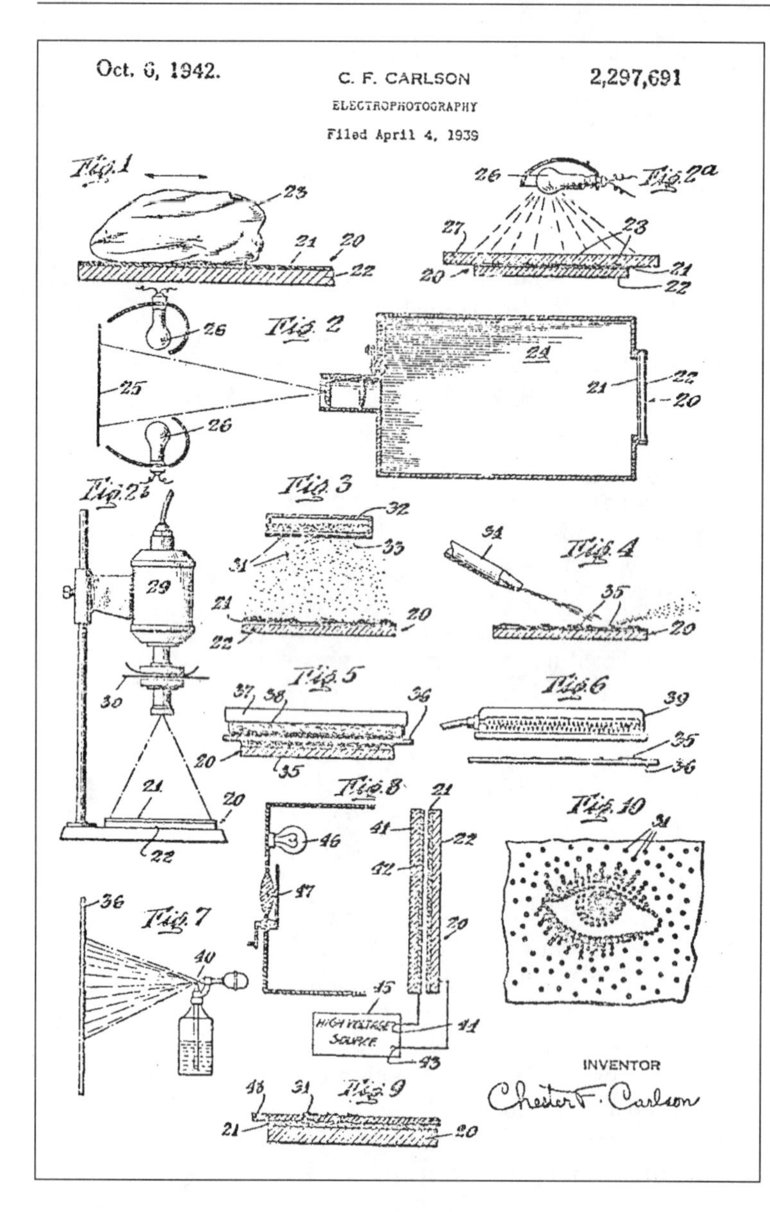

"Dry writing" originally took many steps.

fore introducing the classic Xerox 914 in 1960. Its name came from the maximum size of copy it could make from a roll of paper, 9" x 14". Wildly successful, more than 200,000 of the freezer-sized 650-pound 914s were manufactured. Haloid had expected to make about 4,000. The 914s were never sold. Instead, the company leased each one for about $95 per month plus 5¢ per copy. The 914 could make seven copies per minute. The company's revenues went from $33 million in 1959 to over $500 million in 1966. Briefly called Haloid Xerox, the company changed its name to the Xerox Corp. in 1961.

The holder of 28 patents in xerography, Carlson always had faith in his abilities and never gave up. Though almost penniless in the late 1950s, he later received dozens of awards, including the 1964 Inventor of the Year Award and the 1966 Horatio Alger Award. Carlson had a net worth of $150 million by the late 1960s. He had a weak heart and painful arthritis later in life, and he spent his last years giving away $100 million. Most of his anonymous donations went to research and charities.

References

Xerox World—Special Issue: Chester Carlson's Invention, Fall/Winter 1988.

"Struggling to Become an Inventor" by Dean J. Golembeski, in *American Heritage of Invention and Technology*, Winter 1989.

"Chester F. Carlson, Inventor of Xerography" by Alfred Dinsdale, in *Photographic Science and Engineering*, January/February 1963.

"The Invention Nobody Wanted" by Don Wharton, in *The Kiwanis Magazine*, February 1965.

"Profiles: Xerox Xerox Xerox Xerox" by John Brooks, in *The New Yorker*, 1 April 1967.

Dictionary of American Biography, Charles Scribner's Sons Publishers, 1932; with supplemental updates.

Philo FARNSWORTH

Born:
August 19, 1906, in Beaver, Utah

Died:
March 11, 1971, in Salt Lake City, Utah

At my daughter's wedding shower, which included a light lunch, everybody was impressed with the delicious homemade bread. The bride-to-be approached her seventh-grade teacher and proudly told her that she had made the bread. She said she had found the recipe while researching a related topic for a report in seventh grade, and had made it ever since. The teacher became slightly misty-eyed. She said that teachers often don't know how they may have affected their students. That wedding shower conversation was a minor event, but it has an impressive technical heritage. A teacher at Rigby High School in Idaho spent some extra time in 1922 with a student who had an interest in electronics. Partly because of the teacher's additional effort, the student went on to develop a unique television system.

Smithsonian Institution Photo No. 64079

Early in the history of radio and television, it was not obvious that those media would become huge enterprises controlled by only a few networks. Independent investigators who worked alone helped to launch both communications systems. Many early inventors were left behind in the dynamics of such national enterprises. Lee De Forest's (1873-1961) 1907 triode made broadcast radio a reality. But De Forest earned little recognition or profit from his invention. John Logie Baird (1889-1946) developed the first television system used in broadcast service in 1929. He, too, earned little from his work. Another true pioneer of television is often lost in history's fuzzy memory. Were it not for his work being out of the technical mainstream, Philo Farnsworth might be better remembered as one of the inventors of all-electronic television.

Called Phil by everyone, Farnsworth was born into a south-central Utah farming family. Farnsworth's family had little money and Farnsworth did not learn much about technology until they moved to Idaho in 1920. The house the Farnsworths moved into had back issues of technical publications left behind by the previous residents. Farnsworth read them all several times. Barely in his teens, with no access to related equipment of any kind, he developed an interest in the emerging electronics field. He was especially fascinated by the prospect of using photoelectric cells and cathode-ray tubes to transmit images through the air. Farnsworth was 15 when he began designing his system.

Some television pioneers like Baird were working on mechanical television. This ap-

This early Farnsworth dissector was made from a laboratory beaker. Because it had no scanning beam, it instantly converted an image into electrical pulses and was called a "nonstorage" type of camera tube.

proach used a spinning disk with holes to break an image into small bits. The theory held that such bits could be reestablished at their destination to produce a real-time image. In practice, even the best of the early mechanical television systems produced barely recognizable images. Farnsworth was thinking about an all-electronic system, one that did not use a spinning disk. For technical advice and counseling, he approached Justin Tolman, Rigby High School's chemistry teacher. Farnsworth was new to the town and only a first-year student. Tolman initially thought the young man had an inflated opinion of his ability. But Tolman's opinion soon changed and he freely met with Farnsworth in a large study hall after school.

Farnsworth sketched complicated circuit diagrams on the chalkboard. Then, he and his teacher would evaluate the diagrams and make changes. It wasn't long before the student's knowledge exceeded the teacher's. But it was the technical discussions that encouraged Farnsworth to expand his views of television transmission. Other than use the school library, there was little he could do to obtain current information. The family lived so far from the seats of technical expertise, they were effectively out of contact with the technical world. Farnsworth was in Rigby for just one year before his parents moved again. However, the time he spent there defined the rest of his life.

After graduating from high school in Provo, Utah, Farnsworth started working his way through Brigham Young University. He wanted to learn all about electronics. He needed an educational foundation if he hoped to invent a practical television system. He completed two years of study before his father's death forced him to drop out. Farnsworth was the oldest of five children. He felt a moral and a financial obligation to go to work to keep the family together.

Farnsworth found a job as an office boy with the Community Chest in Salt Lake City in 1926, the year he married Elma Gardner. They eventually had four children. Farnsworth would occasionally discuss his thoughts about television with his co-workers. George Everson and Leslie Gorrell showed considerable interest in his ideas. Neither had any technical knowledge, but they liked Farnsworth's integrity, apparent genius, and work ethic. They invested $6,000 in his work and sent him to Los Angeles, where he rented an apartment not far from the Griffith Park planetarium. He used the dining room as his laboratory. Living in Los Angeles allowed him to meet and communicate with others working in the new field of television. He was barely 20 years old and applied for his first patent within one year. He regularly mailed money home to his family.

The key item in Farnsworth's design was the tube used by the television camera. He named his a *dissector*, because it dissected the image into bits. The photograph on the first page of this chapter shows Farnsworth holding a dissector. It was considerably different from the iconoscope being developed by Vladimir Zworykin (1889-1982) in Camden, New Jersey. Zworykin worked for the Radio Corporation of America (RCA), a giant organization in the radio broadcast field. RCA had a large amount of money invested in Zworykin's iconoscope and did not like the idea of potential competition.

The iconoscope had a scanning beam, while the dissector did not. Farnsworth's dissector was a glass tube that resembled a large can with electrical connections on the side. It had a photosensitive mirror at one end, on which the television image was focused. The mirror emitted electrons in response to the different light intensities projected on it. Farnsworth made the first demonstration of his system during a press conference at his laboratory in 1928. At 150 lines, the images were crude—modern televisions have 625 lines or more. Also, the dissector's electronic efficiency was low, resulting in a dark picture. But it was good enough to receive positive publicity and encourage financial backers. It had taken $60,000 for Farnsworth to reach that stage, a sizable amount at the time. Television development was clearly an expensive proposition.

Farnsworth established a company named Television, Incorporated, in 1929. It was partly funded by the Philco Corporation. His small company would soon compete with large organizations such as RCA, General Electric, and Westinghouse. All had invested large sums of money in the Zworykin system and offered to buy out Farnsworth's company. An intense, high-strung individual, he routinely refused. Always barely keeping his company solvent, Farnsworth made the first open public demonstration of electronic television in 1934 at Philadelphia's Franklin Institute. The image was on a screen only slightly larger than one square foot.

After spending $250,000, Philco dropped its support. In 1935, Farnsworth established his own broadcasting station, W3XPF. But, RCA used its considerable political influence to keep the Federal Communications Commission from implementing commercial television on a nationwide scale. Farnsworth had 165 patents and legally controlled several crucial aspects of television broadcasting.

Developing early television was indeed expensive. By 1939, RCA had spent $9.25 million and Farnsworth about $1 million. No end was in sight and the financial hardships of the Great Depression had taken their toll on Farnsworth. He agreed to sell his patent rights to RCA in 1939. The agreement was a financial triumph for Farnsworth, but it removed him from further active research in television.

Farnsworth then shifted his focus to electronic manufacturing with his Farnsworth Radio and Television Company in Fort Wayne, Indiana. The company built radar equipment during World War II and television sets afterward. It became a division of the International Telephone and Telegraph Corporation in 1958. Farnsworth later did research work with radar and nuclear energy, but he spent much of his time at a fishing retreat in Maine. He died at the age of 64.

It would be encouraging to report that America's large communication networks treated all early-20th-century inventors with an even hand. Yet, they often did not. Many unpleasant company officials were too concerned about corporate profits or a loss of market share. Technical goals were being controlled by business and financial people. Philo Farnsworth had one of America's most inventive minds, but he often faced heavy odds. In the end, he was forced to give up the invention he first formulated as an excited young teenager in the American west.

References

The Story of Television—The Life of Philo T. Farnsworth by George Everson, W. W. Norton and Co. Publishers, 1949.

Electronic Motion Pictures by Albert Abramson, University of California Press, 1955.

"Transition" (obituaries), *Newsweek,* 22 March 1971, p. 76.

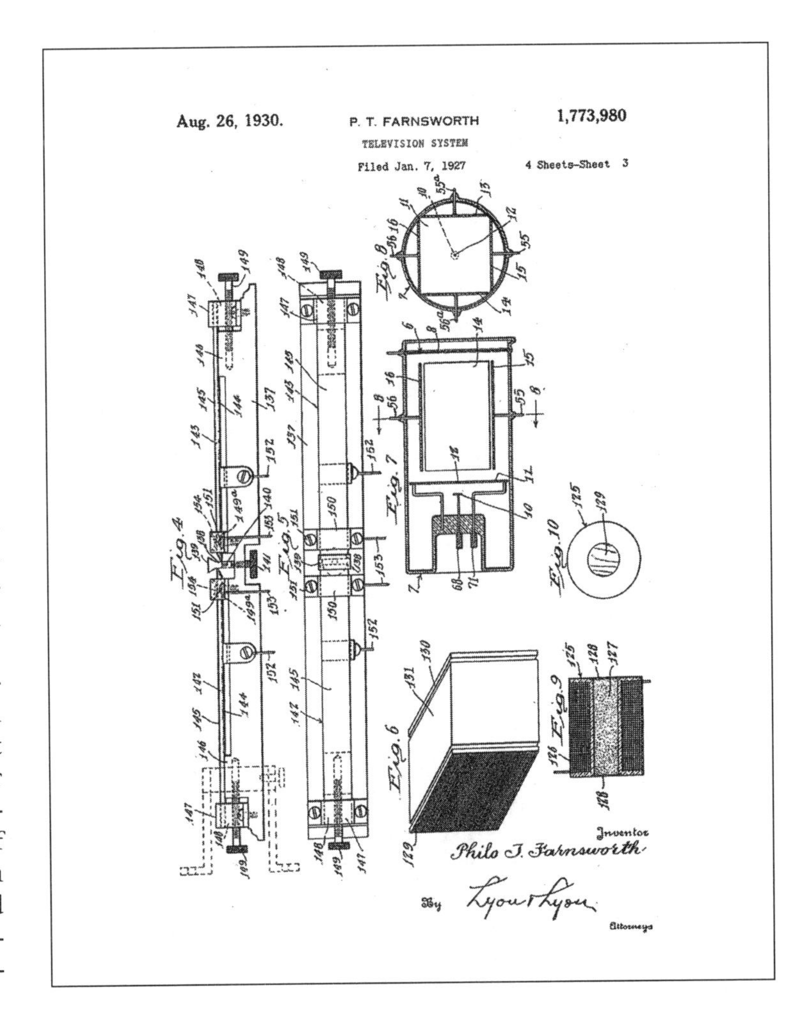

This drawing from a Farnsworth patent shows a cutaway view of his dissector, labeled as "Fig. 7" and "Fig. 8."

Grace HOPPER

Born: December 9, 1906, in New York, New York

Died: January 1, 1992, in Arlington, Virginia

People often say that television has been among the most influential inventions in the world. But that is not precisely true. Television transmitting and receiving equipment does not influence people—programming does. Philo Farnsworth (1906-1971) and Vladimir Zworykin (1889-1982) separately invented different transmission methods. But they were inventors and technologists, not programmers. They were more concerned with developing practical technical systems than in sports, news, movies, or situation comedies. In a similar way, computers can be compared to television. Without programs, a computer would be about as useful as a television set that had no antenna or input signal.

Modern digital computers began to make their tentative appearances in the late 1930s. Before they arrived on the scene, all mathematical machines were externally controlled. Calculators and tabulating machines required routine inputs from operators. Programming was a new idea that suggested a machine with built-in intelligence. The world's first large-scale, automatic, fully programmable computer was the Mark I, developed for International Business Machines by Howard Aiken (1900-1973). It was a huge electromechanical device, more than 50 feet long and 8 feet high. During the dark days of World War II, Aiken knew that the Mark I required accurate programs to make calculations for the U.S. Navy. He enlisted Lieutenant Grace Hopper. She served as one of the first three Mark I programmers. Deciding how to accomplish the new task was a challenging technical activity.

Grace Brewster Murray was born into a middle class New York City family. She was

Smithsonian Institution Negative Number 83-14876

named after her mother's best friend, Grace Brewster. Her father was a disabled insurance salesman who walked slowly on two artificial legs and with the aid of two canes. She said he was her inspiration. But it was her grandfather who cultivated her interest in mathematics. He was a surveyor who allowed Grace to help him on the job. She enjoyed using his colored pencils and developed a fondness for geometry that remained with her all her life. Her father made sure his two daughters and one son had comparable educations and sent Grace to Vassar College in Poughkeepsie, New York. She did not disappoint him and graduated with honors in 1928 with a mathematics degree. She later received two advanced degrees from

Yale University in New Haven, Connecticut. She married Vincent Hopper in 1930.

Vassar College offered her a teaching position in 1931. She remained on the faculty until World War II broke out and she joined the service. Grace Hopper chose the navy because her great-grandfather had been an admiral. She recalled meeting him when she was three. Lieutenant Hopper was assigned to the Bureau of Computation at Harvard University. The bureau used Aiken's huge Mark I computer to calculate trajectories of shells fired from large navy guns. After a brief orientation the first day, Aiken obviously expected big things from Hopper. He said he "would be delighted to have the coefficients for the interpolation of the arc tangents by next Thursday." Hopper joined younger officers Robert Campbell and Richard Bloch. They were the world's first three modern computer programmers.

Hopper was the first person to compile a manual of subroutines. Several small programs, which had been previously checked for errors, could be combined to make an error-free program. Her work was so uniformly impressive that Aiken asked her to write the operating manual for the Mark I. *The Manual of Operation for the Automatic Sequence Controlled Calculator* was the first of its type. IBM was the sponsoring organization and preferred the ASCC name instead of Mark I.

The Mark I was an experimental handmade device in an experimental technology. So it often malfunctioned. Electrical component failure was a common cause, but the Mark I sometimes stopped for unusual reasons. A handwritten "13" was once encoded in the punched tape input as a "B." A symbol for a delta was read as a "4." A later Mark II also had its problems. A moth that had been crushed inside a relay kept the contacts separated. Hopper removed the bug and the computer resumed operating. That is the origin of the term *debug*, which refers to removing errors from a computer program. The moth was taped into the official logbook with the notation "First actual bug found." From then on, if the computer happened to be down, Hopper always said they were "debugging" it.

Hopper stands near a magnetic tape input of a 1960s Remington Rand computer. She helped develop the COBOL book in her hand.

Hopper worked with Aiken for about six years. Then she took a significant career risk and left the navy in 1949. She remained in the navy reserves but went to work as a senior mathematician for a Philadclphia company that later became the Remington-Rand Corporation. Remington-Rand was developing the UNIVAC series of computers, an acronym for Universal Automatic Computer. When unveiled in 1951, the UNIVAC was the first high-capacity computer offered for sale to companies and businesses. The first was purchased by the U.S. Census Bureau. UNIVAC was 14 feet long, had 5,000 vacuum tubes, and was the world's fastest machine. It could use numbers and letters equally well, and it was the first to store data on magnetic tape.

Hopper originated the idea that computer programs should be written in English, instead of complex machine code. The transformation is carried out through a program called a *compiler*. Hopper created the first

The console of an early UNIVAC computer was far more complex than modern desktop computers. The display included the mannequin at the right.

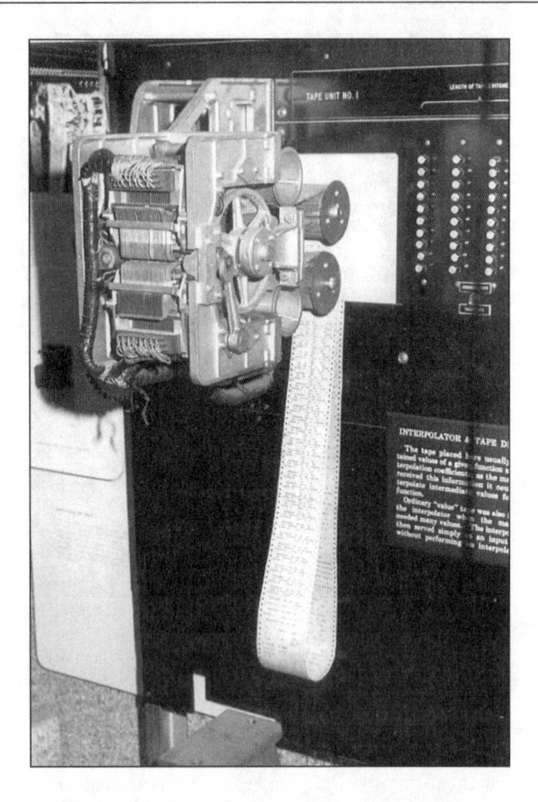

The Mark I computer used four paper-tape inputs like this one. Hopper hand wrote coding commands. An operator used a typewriter-like device to punch holes in paper tape that represented the commands.

one in 1952. It was named A-0 and stored on the UNIVAC's magnetic tape. Hopper knew that the key to opening the field of computers was to develop programming languages that nonmathematical people could understand. Her continued work in that area resulted in Flow-Matic, a language aimed at business applications such as automatic billing and payroll calculations. By the end of 1956, Hopper had Remington-Rand's computers understanding 20 near-English statements. It was just one more step to the development of the enormously popular COBOL (Common Business Oriented Language). Hopper was on the six-person organizing committee that began work in 1959. COBOL was based on her Flow-Matic language.

The navy asked Hopper to rejoin in 1967 to standardize business operations. Its payroll program had been rewritten 823 times and confused everyone. Hopper did what was required but recommended against the use of ever larger computers and punched cards. As long ago as the early 1970s, Hopper favored networked computers. She pointed out that when a farmer wanted to move heavy boulders, he didn't grow a larger ox. He simply added another one.

Hopper's duties often cast her in the role of teacher. Her favorite teaching tool was a wire nanosecond (billionth of a second). A 12-inch piece is about the distance electricity traveled in a nanosecond. Hopper compared it to a 984-foot-long wire microsecond (millionth of a second). She said every programmer should have a wire microsecond so they would know exactly what they were wasting with each extra microsecond.

Hopper's honorary degrees and awards took up eight single-spaced pages in her official navy biography. The most unusual was her election in 1969 as the first ever "Computer Science Man of the Year" from the Data Processing Management Association. And the most significant was probably receiving the 1991 National Medal of Technology from President George H. W. Bush. Hopper was the first woman to receive the award for individual accomplishment. A special appointment from President Ronald Reagan in 1983 advanced her to Rear Admiral, making Hopper the first woman to hold the rank. As an active-duty officer or reservist, Hopper served in the navy for 43 years before retiring in 1986. Her retirement ceremony was held on the USS Constitution, the navy's oldest ship, in Boston Harbor. It was not far from Harvard University where her career had begun more than four decades earlier. Affectionately called Amazing Grace by her subordinates, she never retired from technology. Hopper died quietly at the age of 85, while working as a senior consultant for the Digital Equipment Corporation.

Hopper's wire nanosecond has become something of an icon. But she is also known for a clock that ran backward while still indicating the correct time. She wanted to en-

A moth accidentally flew into this type of relay and caused an early computer to shut down. The relay was "debugged" and the computer returned to operating normally.

courage young people to think in unconventional ways. She felt her greatest accomplishment was the training of countless new computer professionals. Even while in her 80s, Hopper was a charismatic and valued speaker who identified with young people. She often made insightful comments for them to think about. "Go ahead and do it, you can apologize later," was one. Another was, "A ship in port is safe, but that is not what ships are built for." Navy destroyer USS Hopper (DDG 70) was commissioned in her honor on September 6, 1997, in San Francisco.

References

Portraits in Silicon by Robert Slater, MIT Press, 1989.

The Book of Women's Firsts by Phyllis J. Read and Bernard L. Witlieb, Random House Publishers, 1992.

Encyclopedia of Computer Science and Engineering, 2nd Ed., edited by Anthony Ralston, Van Nostrand Reinhold Co., 1983.

USS Hopper's website. Available: http://www.navsea.navy.mil/hopper/

Konrad ZUSE

**Born:
June 22, 1910,
in Berlin,
Germany**

**Died:
December 19,
1995,
in Huenfeld,
Germany**

In the early 1900s, people typically performed multiplication and division with a slide rule. The slide rule is an *analog* device, from the Greek word *analogos*, which means "ratio". Electrical meters with needles and wristwatches with hands are also analog devices. They use the position of needles or hands to indicate numerical values. A slide rule has two identical logarithmic scales that slide over each other. To divide 6 by 3, for example, one number is placed over the other by sliding the scales. The answer, 2, appears where the end of one scale meets the other. Like most calculating devices, the slide rule takes practice, but it's generally easy to use. British mathematician William Oughtred (1575-1660) invented the slide rule in about 1620. American William Burroughs (1855-1898) patented the first keyboard-operated calculator in 1892. Burroughs's device was actually an adding machine. Slide rules were simple, inexpensive, easy to use, and were used to multiply and divide for many years.

However, large and intricate technical projects often required many interim calculations with slide rules. The results were written on pieces of paper for later use. Keeping track of the information often proved a demanding and detailed responsibility. Final answers had to be checked and re-checked. Konrad Zuse, a young German working at his first job in an aircraft factory, had to make many lengthy calculations. He though about the possibility of creating a digital computer. Zuse began building an experimental unit in his parents' kitchen in the late 1930s. He was unaware that others were working on similar projects in America and Great Britain. The world was preparing for war and communication among technologists was often strained.

Courtesy Deutsches Museum, München

Mechanical devices fascinated Zuse when he was a youngster. Early on, he hoped to become an inventor. His father worked for the post office and did what he could to support his son's interest in technology. The younger Zuse studied civil engineering during his college years in Berlin. His class work often required lengthy calculations and he spent time developing a form to keep up with the detailed work. While Zuse didn't know it at the time, he was working on a rudimentary computer program.

Zuse went to work for the Henschel Aircraft Company after graduation in 1935. In his free time, he began building an experimental computer in the apartment he shared with his parents. He named it the Z1. Zuse was proficient with Erector set parts and used them to make his all-mechanical computer. It took three years to complete and was about the size of a four-foot cube. Zuse brilliantly decided to use calculation methods based on binary numbers, 0 and 1. He was the first to determine that using only two characters with his computer would be faster than using the 10 characters in the base 10 number system. That approach is now used in computers throughout the world. Zuse's binary system used rod positions, with one direction indicating "on" and the other "off."

The Z1 had a keyboard input and an output of electric lights. Calculation instructions were punched into discarded 35 mm film and passed through a program reader. Zuse used all these methods in his subsequent computers, as well. He received a patent in 1936. Because the Z1 was made with many lower-quality toy parts, it operated only a few minutes at a time before jamming.

The German army drafted Zuse in 1939 and he served as an ordinary infantryman. Managers at Henschel Aircraft thought his abilities could be put to better use at their factory and they got him back in six months. Zuse returned to the important job of analyzing aircraft wings. The company gave him space and financing to work on his new Z2 computer. Zuse was the only German permitted to develop computers during World War II. He had a proven track record in the field. Government leaders felt that Germany would quickly win the war, and they did not want to make heavy investments in an unnecessary technology. Zuse's was a lonely position. Few friends or colleagues understood what he was doing. Like the Z1, his Z2 was a concept computer and not intended for production use. But his Z3 was meant to be a full-scale operational computer.

The replica Z3 computer is often demonstrated to student groups visiting the Deutsches Museum in Munich. The demonstration is only a simulation because the computer would require too much maintenance to keep it fully operational.

Binary numbers require on and off switching. Vacuum-tube diodes worked well for the purpose because they can switch very rapidly. Zuse suggested a computer with 2,000 tubes, but Henschel was unwilling to pay for such an expensive machine. Instead of tubes, Zuse used 2,600 electromagnetic relays for his 1939 Z3. Their switching rate was a few hundred times per minute. His method paralleled one under development in America by Howard Aiken (1900-1973) with his Mark I computer. Zuse and Aiken did not know of each other's work.

When completed in 1941, the Z3 had cost only $6,500 and it went into service with the German aircraft industry. The three major components included the operator's console, a film reader, and the relays in three cabinets. The Z3's first use was for calculating airplane vibrations during the high stresses of battle. Like the Z1, it used programs punched in 35 mm film. Data was reentered at the console for every new calculation. Operators viewed a series of small lamps to read the binary number output. The Z1, Z2, and Z3 were destroyed in a 1944 air raid.

Zuse supervised the 1962 construction of a Z3 replica for the Deutsches Museum in Munich. Entering data was a precise procedure using numbers and exponents. It followed the form of "A times 2 to the B power." The operator input the proper values for A and B, both of which have to be binary numbers. It was a job requiring great care and

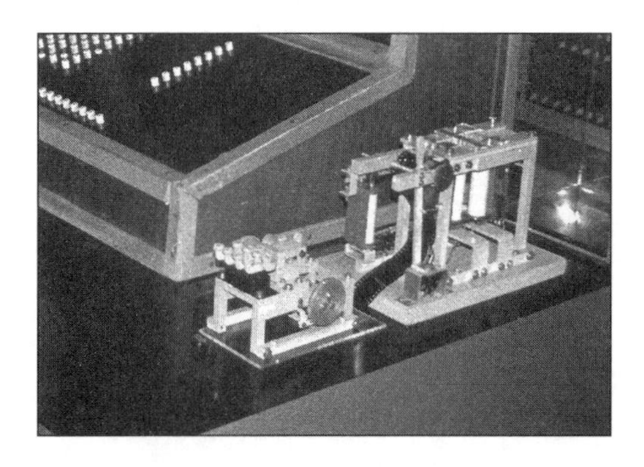

The 35 mm film would normally have punched holes in the center to provide the calculating commands. The film in the reader is for demonstration purposes only.

The replica Z3 computer has three major sections. First, the console contains the input keyboard and the output bank of lights on the vertical section. Second, the program reader at the right passes 35 mm film, which contains the calculating commands. Third, the electromagnetic relays at the back wall provide the on and off switching for binary arithmetic.

attention to detail. But it released the operator from the drudgery of calculations. A 35 mm filmstrip in the reader at the right of the console had the program commands. The 2,600 relays behind the console provided the on and off switching for the binary numbers used in the calculations. The Z3 could add, subtract, multiply, divide, and take square roots. Multiplication took three to five seconds. Data was input through the keypad on the sloped front of the console. The final answer was read from the lamps on the vertical section of the console. There was no paper or screen output.

Zuse quickly completed a slightly upgraded version that he called the Z4. As the war drew to a close in 1945, he loaded the one-ton computer in an old truck. He hid it in the cellar of a farmhouse near the Swiss border. Zuse moved to Zurich in 1945 after the war ended, and the Z4 followed him five years later. It was used by the Federal Polytechnic Institute until 1955.

Zuse established a small computer company in Germany in 1950 and received about 50 patents. He developed one of the first computer languages and named it Plankalkül. It aimed at simplifying mathematical calculations. He sold his company to the massive Siemens corporation in the 1960s and became a consultant to the firm. In his later years, Zuse worked with inventions but spent most of his time doing oil painting. He and his wife, Gisela, had five children. Zuse died at 85 in the western German state of Hesse.

It is not unreasonable to ask who invented the modern digital computer. Most historians bypass Zuse and point to Howard Aiken and his Mark I computer. Reference books have been almost unwavering on this detail. A typical example comes from *The Making of the Micro* by Christopher Evans (1981, page 72): "Though Zuse undoubtedly had a working system operating before the war and Aiken's giant system didn't start working in earnest until 1943, it is the latter system that is generally accepted as having been the first to fulfill [Charles] Babbage's dream."

A related question is, why don't historians credit Zuse's invention as the first modern computer? The answer involves trust. Aiken's computer was clearly operational, had a well-documented history, and survived the war. Zuse's Z1, Z2, and Z3 were presumably destroyed in a 1944 Berlin bombing raid. It is possible they were dedicated calculators and not general-purpose computers. There is little definitive evidence to support Zuse's claim of priority, not even scrap parts following the bombing raid. Unless additional information becomes available, most of the world will credit Aiken's Mark I as the first general-purpose large-scale program-controlled digital computer. However, that hardly detracts from Zuse's innovative, insightful, and pioneering role in computer development.

References

The Making of the Micro by Christopher Evans, Van Nostrand Reinhold Publishers, 1981.

Pioneers of Computing by F. Gareth Ashurst, Frederick Muller Publishers, 1983.

Portraits in Silicon by Robert Slater, MIT Press, 1987.

CONNECTING CIVILIZATION

Index